# Science Mind Stretchers

**by Ann Fisher**

illustrated by Marty Bucella

cover illustration by Marty Bucella

*Publisher*
Instructional Fair • TS Denison
Grand Rapids, Michigan 49544

ISBN: 1-56822-423-0
*Science Mind Stretchers*

2400 Turner Avenue NW
Grand Rapids, Michigan 49544

# Table of Contents

# To the Teacher

Your students will gain a whole new perspective on science instruction when they encounter the activities in *Science Mind Stretchers*. These puzzles are fun alternatives to traditional worksheets, and they are sure to challenge even your brightest budding scientists. While your students are solving these unique puzzlers, they will also be learning and reviewing important science content.

This book is designed to simplify your preparation and instructional time. Each section begins with a teacher resource page. This is a valuable tool which should not be overlooked! The resource pages include tips on what reference material your students will need for the section and what basic knowledge they need before proceeding with the work. Also included are warm-up activities which are often brainstorm/discussion topics that will get students thinking in the "right direction." There are hints for making certain mind stretchers simpler for younger students, and often there are suggestions for extension activities as well. In general, pages are arranged from simplest to most difficult within sections. Fortunately, there is a clear and complete answer key in the back of the book.

A unique feature of this book is that each section ends with two or more pages of open-ended exercises which allow students to think creatively and critically about science topics. Many of the mind stretchers in this book extend beyond the science curriculum into language arts, social studies, and math.

The section "Inventions and Inventors" looks at the history of many important devices and discoveries. "Computers" helps students' understanding of how computers work and what they can do. "Space and Space Exploration" deals not only with important facts about our solar system, but also with the interesting history of space travel. The section "Weather and Climate" teaches about conditions that affect weather and weather-related vocabulary and symbols. "Animals and Plants" exposes students to unfamiliar living organisms and emphasizes categorization. The book ends with "The Human Body," a look at major organs and systems and the marvelous way in which they work.

The next time your students groan when they hear there is a science assignment, surprise them with a fun and challenging page from *Science Mind Stretchers*. You will not be disappointed, and they will not be bored!

# 1 Inventions and Inventors

The study of inventions and inventors is an appealing and valuable topic for students. They can learn a lot about science, history, and research; they can also begin to develop "inventive" ideas of their own. Be certain you have a good selection of appropriate reference books available to students before assigning any of the following puzzlers. Almanacs, encyclopedias, biographical dictionaries, and library books specifically about inventions will also be helpful.

The author would especially like to acknowledge one book which provided much of the information included in this section: *Inventions, Innovations and Discoveries*, by Kevin Desmond, published by Constable and Co. Ltd., London, 1986.

It should be noted that sometimes sources do not agree on exact dates and/or descriptions of early inventions. Facts here have been verified in at least two sources whenever possible.

Warm-up activities:

1. How many famous inventors can your students list on the board in a ten-minute brainstorming session? Copy their list and have each student select one on which to write a one- to two-page report.
2. Write the names of the inventions listed below on the board. Ask students to name the country in which they think each originated. Next, ask them to place them in chronological order, numbering them from earliest to latest. (Answers in parentheses.)

gunpowder (China) (2; 221 BC)
boxing gloves (England) (7; 1747)
CD-ROM (Japan) (10; 1985)
windmill (Persia) (3; 644)
bathtub (Babylonia) (1; 1800 BC)
armored ship (Korea) (5; 1592)
satellite (USSR) (9; 1957)
printing press (Germany) (4; 1451)
telephone (USA) (8; 1876)
classified advertisement (France) (6; 1631)

## Suggestions for Specific Puzzlers

### Chinese Confusion

This is a fun and informative activity in which disaster could result if instructions are not followed! Advise students to read and proceed carefully.

### Criss-Crossed Inventors

Have reference books available for this activity. Inventors named here can be added to the list begun in the warm-up section to be assigned as report topics.

**Sequential Sets**
This activity could be assigned to small groups. Ask each group to first agree on the earliest invention, using students' own knowledge and guesswork. Then, allow the group to divide up the research, working cooperatively to find as many dates as possible. Note that some dates may vary from book to book.

**International Inventiveness**
If students have difficulty, you may wish to supply the first letter of some of the answer words. Or, if students have solved one word in a set, you may choose to allow them to look up its country and fill in the circled letters before unscrambling the other items.

**Enterprising Edison**
See how many questions students can answer on their own before allowing the use of biographies, encyclopedias, etc. Or, use this in small groups so that, without reference books, students have to pool their knowledge (and/or guesswork) to settle on their best answers. Be certain, of course, to cover the correct answers in the end.

**Scientific Match-Up**
This straightforward activity will no doubt require the use of reference books.

**"Like"-ly Inventions**
Here again, students will need reference material. For some sets, answers may not be apparent until all three items are researched.

**Preposterous Publicity**
This activity can be used in the following ways: (1) Read paragraphs orally and have students jot down their best guess. (2) Assign the page to individuals so that they have to research answers. (3) Assign to small groups so that each group either divides up the research or has to agree on their "best guess" for the answer.

**True Tales**
Using only the context (and perhaps some lucky guesswork), students should be able to complete most answers. Use reference books only as a last resort.

**Before, What Is It?** and **It's Your Turn**
These are all open-ended activities designed to let your students indulge in their own creative thinking. Allow students to share outcomes, as desired, with their classmates. Feel free to extend these in any appropriate manner.

Name ____________________

# Chinese Confusion

You may already know that China was the birthplace of rockets, paper money, and gun powder. But there is another great (and very useful) invention for which we can thank the Chinese. To find out what it is, carefully follow the instructions below, using the numbers and letters in the boxes. When you cross out a number, also cross out the letter that appears underneath it.

| 1 | 2 | 3 | 4 | 5 | 6 | 7 | 8 | 9 | 10 | 11 | 12 | 13 | 14 | 15 | 16 | 17 | 18 | 19 | 20 |
|---|---|---|---|---|---|---|---|---|---|---|---|---|---|---|---|---|---|---|---|
| A | H | S | G | J | D | U | R | E | C | B | H | T | L | M | O | O | N | T | Y |

1. Read all the directions before doing anything.
2. The wheelbarrow was invented in China in the year 231. Cross out the number that shows how many wheels are on a wheelbarrow.
3. Cross out the number that is the sum of the digits in the year the wheelbarrow was invented.
4. The Chinese invented rockets in 1100. Cross out any number above that divides evenly into 1100.
5. Porcelain was invented by the Chinese in 700. Count the number of letters in the word PORCELAIN. Cross out that number and all its multiples.
6. In 580, the Chinese invented the suspension bridge. Cross out any letters remaining above in the word BRIDGE.
7. Count the letters in CHINA. Cross out that number and all its multiples.
8. In the year 400, the Chinese invented a ship with watertight compartments. Cross out any numbers above containing the digit 4.
9. In 400 B.C., the kite was invented in China. Cross out any letters above that appear in KITE.
10. Reverse the leftover letters and write them here.

    ___ ___ ___ ___ ___ ___ ___ ___ ___ ___

11. Do not do any of the instructions in number 4, 6, or 9.

Now guess what year the item above was invented. ______________

Name ______________________________

# Criss-Crossed Inventors

Each miniature crossword provides the name of an invention. The blanks in each set are where you are to complete the last name of the inventor or discoverer. Use your own knowledge and reference books to complete each missing name. For example, the inventor of the motion picture in number three was Edison. Fill in the missing letters D, I, S, and N.

Name ______________________________

# Sequential Sets

Below are sets of related inventions. For each category, first circle the item that you believe was invented the earliest. Then research as many dates as possible and write them in the blanks below each item. How many of your predictions were correct?

1. Automation:
   A. milking machine ______ B. spinning jenny ______ C. McCormick's reaping machine ______
2. Clothing:
   A. mini skirt ______ B. blue jeans ______ C. dinner jacket ______
3. Medicine:
   A. aspirin ______ B. tuberculosis vaccine ______ C. public hospital ______
4. Architecture:
   A. stained glass window ______ B. Eiffel tower ______ C. Brooklyn Bridge ______
5. Musical instruments:
   A. harp ______ B. clarinet ______ C. steam organ ______
6. Correspondence:
   A. adhesive postage stamp ______ B. postcard ______ C. fountain pen ______
7. Water transportation:
   A. commercial steamboat ______ B. canal locks ______ C. iron boat ______
8. Foods:
   A. tea bags ______ B. canned baked beans ______ C. croissant ______
9. Technology:
   A. word processor ______ B. video game ______ C. bar code system ______
10. Scientific discoveries:
    A. atomic theory ______ B. magnetic field ______ C. gravity ______
11. Entertainment:
    A. phonograph ______ B. public radio broadcast ______ C. juke box ______
12. Consumer conveniences:
    A. drive-in bank ______ B. self-service grocery ______ C. department store ______

Name ______________________

# International Inventiveness

Individual countries take great pride in being the birthplace of unique inventions and important discoveries. Unscramble the name of each invention below, and write it in the blanks provided. When you are done, the name of a nation will appear in the rectangles—the nation of origin for each item in the group.

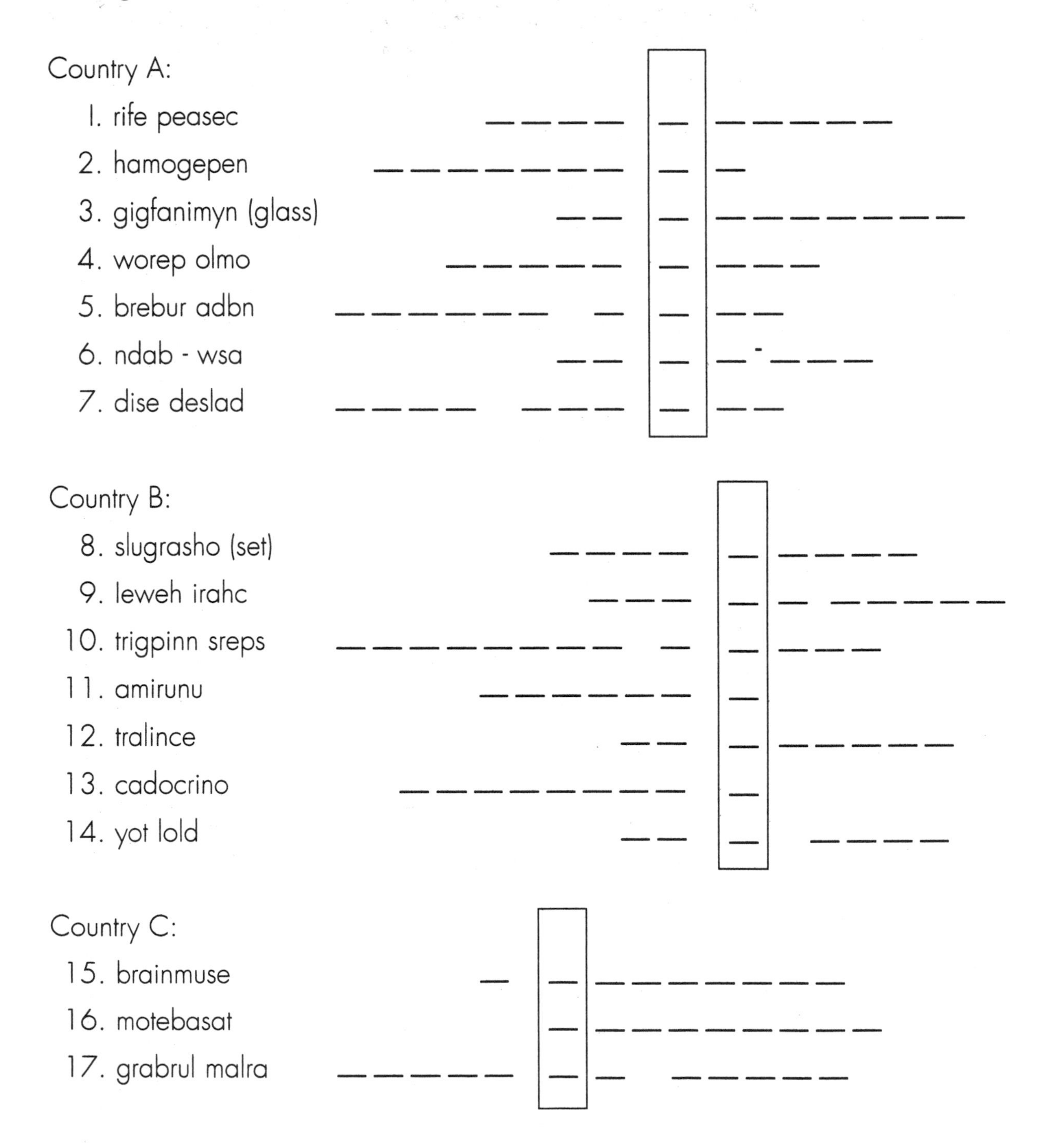

Name ______________________________

# Enterprising Edison

No collection of invention activities would be complete without a feature on the world's most prolific inventor, Thomas Alva Edison. Do your best to find the right answer to each question below. Circle the letter of the correct answer(s).

1. During his lifetime, Edison took out no fewer than __________ patents.
   A. 100 B. 500 C. 1000

2. Thomas Edison was born in the year __________.
   A. 1787 B. 1847 C. 1907

3. He was born in the city of __________.
   A. Dearborn, Michigan B. Milan, Ohio C. Trenton, New Jersey

4. Thomas Edison attended school for __________.
   A. 3 months B. 3 years C. 6 years

5. He left school early because:
   A. He was needed to help on his family's farm.
   B. He became seriously ill.
   C. His teacher told him he was not smart enough to be in school.

6. Tom's first experiment was:
   A. At age 4 he tried to hatch goose eggs.
   B. At age 8 he tried to improve the telegraph.
   C. At age 10 he tried to convert water power to electricity.

7. This first experiment was
   A. a success. B. a failure.

8. Tom's first job was
   A. being a delivery boy for a grocer.
   B. milking cows on a neighbor's farm.
   C. being a newsboy on the train.

9. To increase his earnings on this first job, he decided to
   A. print his own newspaper to sell.
   B. bake his own bread to sell.
   C. hire others to help him, paying them a lower. wage than what he received.

10. During his time at this first job, Thomas had a small accident resulting in a life-long injury to his
    A. vision B. hearing C. left foot

11. During the Civil War, Thomas worked as a
    A. messenger for the Army B. railway engineer C.telegraph operator

12. After working on inventions for other men for about eight years, at age 29 Edison bought a place where he could work on inventions of his own. This was called:
    A. Gray Square in Iowa B. Think Place in New York
    C. Menlo Park in New Jersey

13. Some of Edison's inventions include (circle all that are correct):
    A. electric vote-recording machine
    B. megaphone for use by the deaf
    C. telephone switchboard
    D. penny farthing bicycle
    E. talking motion pictures
    F. alkaline storage battery

14. The first message recorded on another important invention, the phonograph, was:
    A. "I did it!"
    B. "Twinkle, twinkle little star. . ."
    C. "Mary had a little lamb. . ."

15. One of Edison's greatest inventions was the incandescent light bulb. Other light bulbs had been invented earlier, but his was an improvement (particularly over the carbon arc light) because it (circle all that apply):
    A. was made of plastic D. produced a long, steady light
    B. was cleaner E. was safer
    C. was inexpensive F. produced a colored light

16. When Thomas Edison was 67, his second lab in West Orange, New Jersey, was destroyed by fire. Edison decided to
    A. retire B. keep on inventing C. move to Arizona

17. After an illness at age 84, Thomas Edison died. His wife signalled reporters waiting outside the house by:
    A. playing music on the phonograph
    B. closing the drapes
    C. turning out the light in his bedroom

Name ______________________

# Scientific Match-Up

Match the originator and year with the correct scientific discovery by writing a letter in each blank. Complete as much as you can without any aids; then refer to research material to find the remaining answers.

| | | |
|---|---|---|
| ____ | 1. China, A.D. 80 | A. Theory of gravity |
| ____ | 2. Kepler, 1611 | B. Temperature scale |
| ____ | 3. Boyle, 1662 | C. Space shuttle |
| ____ | 4. Newton, 1684 | D. Black holes |
| ____ | 5. Romer, 1690 | E. Theory of relativity |
| ____ | 6. Fahrenheit, 1715; Celsius, 1742 | F. Rainbow theory |
| ____ | 7. Black, 1756 | G. Uranus |
| ____ | 8. Herschel, 1781 | H. Atomic theory |
| ____ | 9. Klaproth, 1789 | I. Carbon dioxide |
| ____ | 10. Dalton, 1803 | J. Magnetism |
| ____ | 11. Oersted, 1819 | K. Robotics |
| ____ | 12. Wohler, 1827 | L. Magnetic north pole |
| ____ | 13. Ross, 1831 | M. Gas pressure laws |
| ____ | 14. Einstein, 1905 | N. Pluto |
| ____ | 15. Tombaugh, 1930 | O. Magnetic field |
| ____ | 16. NASA, 1960 | P. Speed of light |
| ____ | 17. Rand Corp. and IBM, 1962 | Q. Weather satellite |
| ____ | 18. IBM, 1970 | R. Computer floppy disk |
| ____ | 19. Boyd, 1972 | S. Uranium |
| ____ | 20. NASA, 1977 | T. Aluminum |

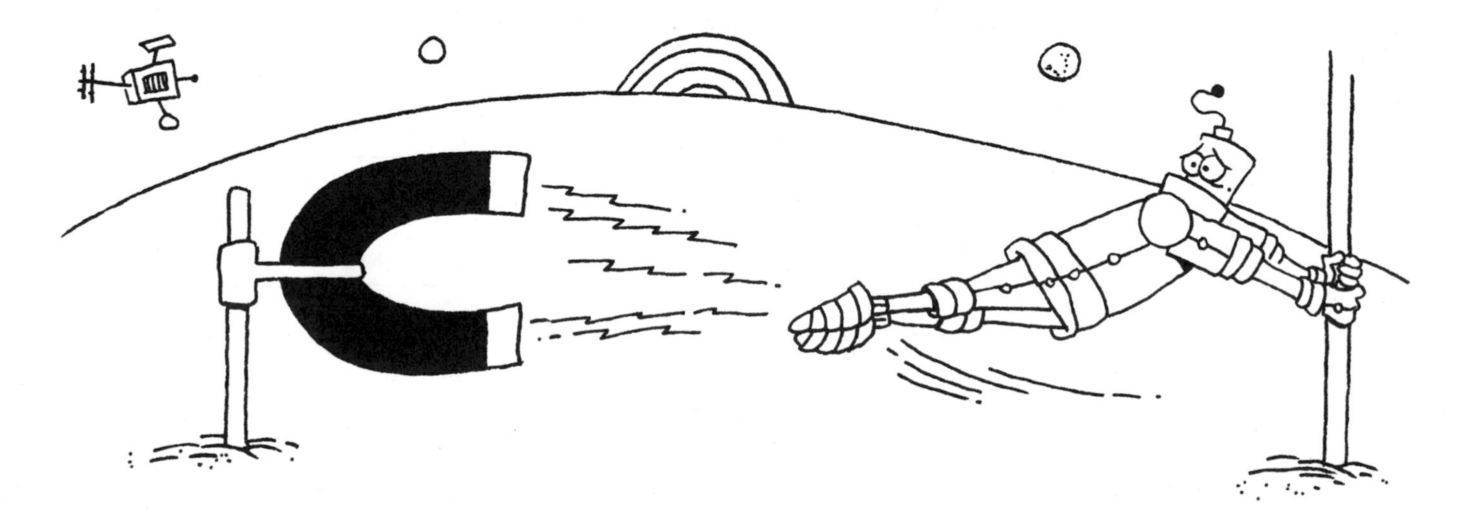

Name ____________________

# "Like"-ly Inventions

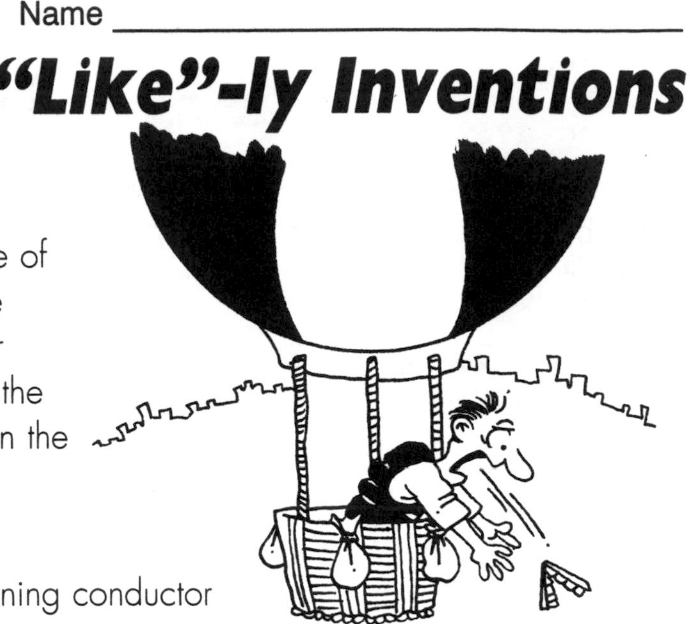

Each set of items below has something in common. The items may be alike because of their inventor, the year in which they were developed, their country of origin, or their general purpose. When you find why all the items in a set are alike, write the reason in the blank provided.

1. bifocal lenses, glass harmonica, lightning conductor

   ____________________

2. cellophane, filter coffee machine, paper drinking cups

   ____________________

3. drive-in bank, shopping cart, photosensitive glass

   ____________________

4. graphite-clay pencil, paper-making, typewriter

   ____________________

5. ambulance, sonnet, medical thermometer

   ____________________

6. Richter magnitude scale, Model T Ford car, Graham cracker

   ____________________

7. wireless telegraphy, escalator, helium

   ____________________

8. horse bit, cotton gin, artificial fertilizer

   ____________________

9. life jacket, false teeth, hot-air balloon

   ____________________

Name ___________________________

# Preposterous Publicity

These news stories could not possibly be accurate, because, on the date given, one of the items mentioned was not yet invented. In the blank below each story, write the item that is out of place, along with its date of origin, if possible.

A. Tuner Trips: June 13, 1765

Last Thursday, Mr. Keith Fisher, local piano tuner, was injured at the home of a client, Mrs. Diana Nixon. Mr. Fisher completed the tuning and put his tuning fork away. Since it was a sweltering 88° F, Mrs. Nixon then offered the tuner a cold glass of fizzy mineral water. As Mr. Fisher reached for the glass, he slipped on little Timmy's roller skates, sending the water across the jigsaw puzzle on the table and sending Mr. Fisher's forehead into Mrs. Nixon's metal sewing scissors. After Dr. Blechl arrived and treated Mr. Fisher's wound, Mrs. Nixon fed both the tuner and Timmy a sandwich and sent the boy to kindergarten.

Answer: ____________________ Correct year____________________

B. Bank Break-In: April 25, 1950

A burglary was reported at the city's All-Saver's Drive-In Bank last night. Billy Jones, who was riding down Hill Street on his skateboard, reported seeing a man in a dark-hooded coat running from the building about 6 p.m. Police officers said the burglar gained entrance to the bank when he threw a can of window polish through the glass pane on the front door and reached through the frame and unlocked the handle. Bank officials stated that all money was secured in the vault. The only items reported missing were these from the bank president's office: an electric guitar left by his teenage daughter and an espresso coffee machine, which had been a Christmas present from his wife. Police continue to look for suspects who may have known the family and were aware that these items were being kept in the bank office.

Answer: ____________________ Correct year____________________

Name ______________________________

# True Tales

Below are ten brief but true tales about some interesting inventions. Your job is to simply fill in the blank with the correct invention or a characteristic of an invention. Some answers will be obvious. Others will require deeper thinking and even speculation.

1. A china shop owner named Melville Bissell of Grand Rapids, Michigan, suffered from terrible headaches. He became convinced that the cause of his headaches was the dusty straw in which the china was packed. In 1876, he developed a ________.

2. In 1927, the Rolex Company of Switzerland invented a watch called the "oyster." It was unique because it was ______________________.

3. John Spilsbury of London invented the ______________ ______________ in 1763 to assist in the teaching of geography.

4. Air pressure outside and suction from indoor elevators made doors difficult to open in new skyscrapers. So, in 1888, Theophilus van Kannel from the United States invented the ______________ ______________.

5. In Holland in 1683, Anton van Leeuwenhoek observed little animals which he called "animolcules." He used a high-powered lens to find these in saliva, teeth scrapings, cow excrement, etc. Unknowingly, he had discovered ______________________.

6. Lt. Walter Wilson and William Tritton of the United Kingdom made a prototype of a tracked, armored vehicle in 1916. It was so successful that 100 vehicles were ordered. For security reasons, these were called "water carriers." Today they are commonly referred to as______________.

7. The Emperor Nero in A.D. 60 used the large transparent gemstone on one of his rings as the first miniature______________________ ______________________.

8. In 1734, M. Fuchs of Germany filled glass balls with water and threw them into a fire, thus inventing the first ____________________ ____________________.

9. Frenchman Dinis Papin designed a container in 1680 with a tightly fitting lid which increased the pressure inside and raised the boiling point of water. Papin claimed his "bone-digester" would make the holdes, hard pieces of beef, tender and savory. Today his invention is known as a ________________ ________________.

10. In 1905, the Automatic Hook and Eye Co. of Hoboken, New Jersey, began machine-producing a type of closure under the trade name of "C-Curity." It is better known today as a ________________________.

Name ______________________________

# Before

Do you ever wonder how people lived before today's modern conveniences—both major and minor? Here is a chance to think and write about a few specific circumstances. Be as practical or as creative as you like as you answer the following questions. Write a paragraph for each question on another sheet of paper.

1. The toothbrush was invented in 1498, and the toothpaste tube in 1892. Do you think toothpaste was used before 1892? If so, how was it sold, handled, and used? What was toothpaste like back then?

2. On another dental topic, forceps used for extracting teeth were invented in 1525. What do you think happened before that? Were teeth not pulled? Or were they pulled with something else? What?

3. Printed books were used as early as A.D. 868 in China. But page numbers in printed books were not used until 1470. What problems would that present? How did people refer to specific book excerpts without page numbers?

4. Numbers were used thousands of years ago—long before the time of Christ. But mathematical signs came much later: + and - in 1489; = in 1557; x in 1631, and so on. How did people write math sentences before then? What difficulties might they have had?

5. Graphite pencils were introduced in 1565. Erasers made of India rubber were invented in 1770. In between those years, how were mistakes corrected? What materials might have been used?

6. The stapler was invented in 1868 and the paper clip in 1900. Before then, how did people keep their papers organized? How might they have attached several pages together?

7. Theater curtains were first used in 1664 in Japan. How would theatrical performances have been different before then, without the curtains?

Name ______________________________

# What Is It?

This is your latest invention. Write its name in the blank provided. Modify the drawing, if necessary, and color it to make it just right. In the lines under the drawing, give a brief description of how your invention works and what it does.

______________________

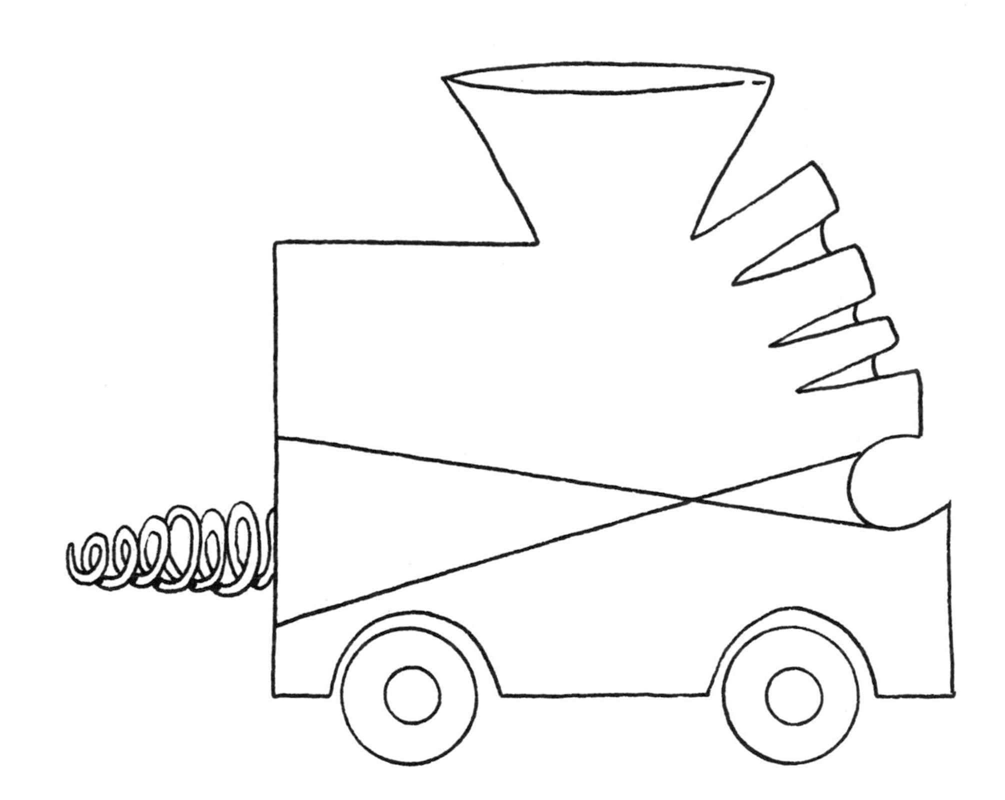

______________________________________________________________

______________________________________________________________

______________________________________________________________

______________________________________________________________

______________________________________________________________

______________________________________________________________

Name ______________________________

# *It's Your Turn*

The world is just waiting for a brilliant new invention—from you! Here are some bizarre ideas to help you get started. Work on a few that interest you and then focus on your best idea. Write a one-page description of your new invention. Include information on your item's purpose, construction, marketability, and cost. Also include a drawing and a sample magazine advertisement for your new invention.

Sample Ideas for New Inventions:

1. a futuristic energy-efficient building for work, recreation, or living
2. a safe and fascinating new toy for babies
3. a hat for people with unruly hair
4. another way to construct books
5. the latest comfort inside a car
6. a re-invented wheel
7. a unique transportation device
8. the all-improved hamster cage
9. a new form of communication
10. a gadget to assist your teacher in all the daily work
11. a musical instrument
12. a new kitchen tool for chopping, slicing, or dicing
13. a new method for building roads
14. all-purpose shoes
15. a new piece of exercise equipment
16. a new idea in home decorating
17. luggage everyone will want to take on their next vacation
18. the necktie that does more

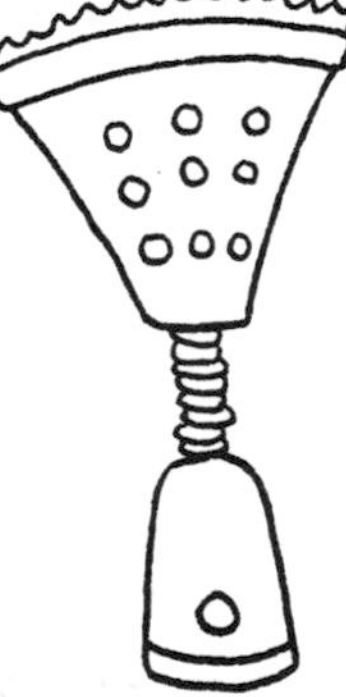

# 2 Computers

This section will help to increase student understanding of what a computer really is, what it can do, and how it works. Most students have used computers, but many may have very little understanding of how they work. Exercises are intended to be basic enough for beginners to understand; the "puzzling" element, however, will challenge all students.

NOTE: Exercises regarding binary codes, databases, programs, etc., are not intended to be technically precise or exhaustive. Rather, their purpose is to increase student understanding of organizational and logical skills needed in using them.

While it is useful to have a computer and some computer manuals on hand for this unit, it is not essential. These mind stretchers have been written so that anyone can work on them in any classroom. Reference books on computer basics will be useful.

Warm-up activities:

1. Ask the class to define a computer. Some simple definitions include "an information processor" and "a machine which does things with information."
2. Discuss in which situations a computer is helpful and in which it is not.
3. As a class, list ten jobs a computer can do quickly. (Calculations, alphabetizing, sorting, correcting spelling errors, and playing games are examples.)
4. Have students draw their own replica of a computer keyboard. This will introduce them to placement of letters, numbers, symbols, and functions.

## *Suggestions for Specific Puzzlers*

### Humanity vs. Technology

Answers provided in the key are the most obvious; however, be ready to accept other answers which can be satisfactorily explained by students. (For example, on number 6, some might argue that their mothers can talk on the phone, while typing on the computer, while cooking dinner!) In general, students should conclude that computers are best at speed, handling large quantities, memory, multi-tasking, precision, predictability, and reliability. Humans are better at flexibility, adaptability, creativity, and perceptiveness.

### Basic Match-up

This may be too basic for some levels, but for others it can provide a fundamental understanding of what different computers are and how they are used.

### Device Dilemma

This may look very difficult to solve, but encourage students to take it one step at a time. By filling in all the letters P and R, they should be able to guess the word PRINTER. That then reveals all other letters of I, N, T, and E. Next, they may be able to guess PLOTTER, which then leads them to ICONS, SCREEN, and so on.

**Chip, Chip Hooray!**

Ask students to read the steps very carefully; there is enough information presented for them to figure this out logically.

**Definite Mix-Up**

Accept reasonable variations in the answers, being certain that all exact words are used. Students may want to write more mixed-up definitions of their own. Possible terms (some of which are defined on other pages in this unit) include *data, desktop publishing, input, microprocessor, output, printer, program,* and *programming language.*

**Binary Baffler**

If necessary, cover exponents and powers of ten and two more thoroughly. As students complete Part II, they should begin to see patterns and, when they do, they may be able to complete it quite quickly. As a follow up, give students more numbers to convert to base two and more words to decode or encode for a partner.

The rest of the Mind Stretchers in this section begin to ask students for more open-ended answers. Accept any that seem reasonable. The goal here is to get students to THINK logically through many steps needed in working with common computer programs.

**Name That Document!**

This is a fun exercise in deciphering shortened titles and descriptions. Accept all reasonable answers. You may want to share outcomes as a class to see the wide variety of answers that are appropriate.

**Field Finesse**

The hardest part of working with any database is setting it up. One has to consider every conceivable way in which it may be used in the future. Students are asked to think through which fields would be needed for a wide variety of topics, which takes them beyond the simple input-output exercises into critical thinking skills.

**Pumpkin Program Problems**

This provides a most basic look at computer programming. If your students have no previous experience, this will be an excellent activity to concisely explain the process and take them quickly into writing programs. If your students are more advanced than this, use the same format to suggest more complicated situations. Be certain that they draw a flowchart first and watch for "bugs" in their work.

**Keyboard Quest**

This is a fun way to practice finding the locations of letter keys. You may wish to allow students to work in pairs; you may also wish to have students share their outcomes with the class. For example, find the longest word formed with the home-row keys, etc.

Name ______________________________

# Humanity vs. Technology

Think about the differences in the ways humans and computers perform tasks. Although computers can greatly simplify our daily lives, some things are still done best by humans. Consider why that is true as you complete this page.

A. Here are ten descriptions of how tasks might be performed. Some best describe computers; place a C in the blanks in front of those statements. For items that best describe humans, place an H in the blanks.

____ 1. It works consistently with very large amounts of information.

____ 2. It can adapt new information to the present situation.

____ 3. It performs the same task in slightly different ways each time.

____ 4. It can recall every piece of information it has ever received.

____ 5. Partway through a task, it can change its mind and proceed another way.

____ 6. It can perform many different tasks at the same time.

____ 7. It can select when it wants to work on which task.

____ 8. It needs a small amount of space to store a lot of information.

____ 9. It is predictable and reliable.

____ 10. It can create a new outcome.

B. Now write four additional descriptions. Write two that best fit computers and two that tell about humans.

Computers:

11. ______________________________

12. ______________________________

Humans:

13. ______________________________

14. ______________________________

Name ______________________________

# Basic Match-Up

Check your knowledge of some computer basics as you complete this page. You may already know that there are several types of computers. For each kind of computer listed here, write the letter of its description and the number of its picture in the blank.

_____ 1. Mainframe

A. Smaller than a mainframe but too large for a desk. Generally used to do one job. May have many terminals linked to it.

I.

II.

_____ 2. Minicomputer

B. Small and inexpensive enough to be owned by many people. Often called a PC. Can run a variety of programs and use many input and output devices

_____ 3. Microcomputer

C. Portable with built-in screen, storage, and power supply. Can be connected to printer or another computer.

III.

_____ 4. Computer Network

D. So large and powerful that it may fill rooms. Used to store and process vast amounts of data. Has many terminals linked to it that do not have their own memory and processor.

IV.

_____ 5. Lap-top Computer

E. Terminals with their own memory and processing units are linked together to additional central memory processing units. Enables one to use and share data.

V.

Name ______________________________

# Device Dilemma

A computer is given information (data) and a set of instructions (program) that tells it what to do with the information. The data that goes in is *input*. The results which come out are known as the *output*. Most commonly, the computer uses a keyboard for input, but there are many other ways of getting information in and other ways of getting it out. Can you think of some?

Listed below are ten devices written in code. Each letter represents another letter of the alphabet. The letters that can be uncoded for G, P, and R are given. First, replace every H you see with an R, every L with a P, and every O with a G. Then try to uncode the rest of the letters to uncover the ten input or output devices.

A =
B =
C =
D =
E =
F =
G =
H = R
J =
K =
L = P
M =
N =
O = G
P =
Q =
R =
W =
Z =

1. J F H  Z W P M A
______________________ ____

2. L H Q K G M H
______________________ ____

3. A Z F K K M H
______________________ ____

4. L D W G G M H
______________________ ____

5. Q Z W K A
______________________ ____

6. R W C A M
______________________ ____

7. R C A Q Z F D  E M N J W F H P
______________________ ____

8. P Q A E
______________________ ____

9. A Z H M M K
______________________ ____

10. O H F L B Q Z A  G F J D M G
______________________ ____

Now go back over your list of decoded words. Place an I in the blank by the ones that are input devices. Write an O by the ones that are for output. If you are uncertain about any, try to look them up in a manual or dictionary.

Name ______________________________

# Chip, Chip Hooray!

Electrical pulses do all the work inside a computer. The pulses are controlled by electronic components. In the first computers, these components were glass valves, which used a lot of power, grew very hot, and were unreliable. Next came the smaller and cheaper transistors. They used much less power and stayed cooler.

But, in the 1960s, the development of the silicon chip revolutionized the construction of computers. The integrated circuit, or "chip," is a tiny slice of silicon on which millions of components are packed closely together. With the chip, it is now possible to build smaller, faster computers than ever before.

To find out how these amazing chips are made, place the steps below in chronological order. If you read the steps carefully and use some logic, you should be able to sequence them correctly. Number the steps from I to 10 in the order in which they must be performed.

____ 1. The chips are then tested to see if an electric current can pass through each circuit.

____ 2. The silicon is sliced into thin wafers. Later, up to 500 chips will be made from each wafer.

____ 3. The silicon wafers are placed in a furnace at a temperature of over 1800°F.

____ 4. The different circuits are built up, one on top of the other, in the silicon wafer. The circuits are designed using a computer and drawn up to 250 times larger than they will be on the finished chip.

____ 5. After being cut, the chips are again inspected. Faulty ones are thrown away.

____ 6. Crystals of pure silicon are grown in a vacuum oven. The silicon is so pure that it will not conduct electricity until treated with certain chemicals.

____ 7. The wafer is cut into individual chips by a diamond or a laser saw.

____ 8. Each tiny chip is placed inside a plastic case with gold wires connecting the circuits to the pins in the case.

____ 9. In the intense heat of the furnace, atoms of certain chemicals enter the surface of the silicon, along the printed lines of the circuit.

____ 10. The circuit designs are then reduced to the size of the chip and printed one at a time onto the silicon wafer.

Name ______________________________

# Definite Mix-Up

Below are several important computer terms and their mixed-up definitions. All the words in the definition are present, but they are out of order. Rewrite the definitions correctly on the lines provided. Consult computer manuals or dictionaries, if necessary, to help you.

1. BINARY - based 0 system on number a digits: two 1 and

______________________________

2. BIT - is or digit, 1 this 0 binary a

______________________________

3. BUG - a program computer in error an

______________________________

4. BYTE - binary represents memory a unit of of group digits one eight data the in which computer's

______________________________

______________________________

5. CPU - (Central Processing Unit) - inside the which all parts computer center of control the organizes other the

______________________________

______________________________

6. DATABASE - can accessed stores in program such that way a information quickly that a it be very

______________________________

______________________________

7. DISK - data round on magnetic stored is a flat which plate computer

______________________________

8. FLOWCHART - showing needed program for steps a computer chart a the of sequence

______________________________

______________________________

9. HARDWARE - computer all itself equipment the and devices input, the including output, storage computer

______________________________

______________________________

10. KILOBYTE (often called "K") - measure equal storage a data bytes of 1,024 to

---

11. MEGABYTE - measure storage of data equal a bytes to 1,024 x 1,024

---

12. MEMORY - instructions the code information chips where and the stored computer in are binary in

---

---

13. MODEM - lines enables communicate a which to telephone with device computers other each using

---

---

14. MOUSE - roll input desk pointer commands an input move on-screen device you to which to around a on an

---

---

15. RAM (Random Access Memory) - memory and stored part of are computer's the results temporarily the where instructions, data,

---

---

16. ROM (Read Only Memory) - permanent containing the a memory computer's part of instructions of the storage

---

---

17. SOFTWARE - disk programs or computer tape on

---

18. VIRUS - secretly, computer cause software a but problems the deliberately, bug to introduced a for user or to

---

---

Name ______________________________

# Binary Baffler

As you already know, computers use a binary code. This means they use only these two signals: *pulse* which is "on" or "1" and *no-pulse* which is "off" or "0." The code of pulses is created by the transistors in the chips which act like switches, turning the current on and off. These series can represent anything—numbers, letters, or even colors.

To see how pulses can represent numbers, look at base two numerals. In base two, each written digit is either a 0 or a l. Place values increase by powers of two, whereas in the decimal system we commonly use, place values increase by powers of ten. Compare:

Base 10: 4,269 = 4 groups of 1,000 or $4 \times 10^3$ = 4,000
plus 2 groups of 100 or $1 \times 10^2$ = 200
plus 6 groups of 10 or $6 \times 10^1$ = 60
plus 9 groups of 1 or $10^0$ = 9
4,269

Base 2: 1,011 = 1 group of 8 or $2^3$ = 8
plus 0 groups of 4 or $2^2$ = 0
plus 1 group of 2 or $2^1$ = 2
plus 1 group of l or $2^0$ = 1
11

So the number 1011 in binary code (or base two) is equal to 11 in our decimal system.

I. Study these numbers that are already in a place-value chart. Write their base 10 equivalent.

| | $2^5$ (32) | $2^4$ (16) | $2^3$ (8) | $2^2$ (4) | $2^1$ (2) | $2^0$ (1) | | |
|---|---|---|---|---|---|---|---|---|
| A. | | 1 | 1 | 0 | 0 | 1 | = | ______ |
| B. | | | | 1 | 1 | 0 | = | ______ |
| C. | | 1 | 0 | 1 | 1 | 0 | = | ______ |
| D. | | | 1 | 1 | 1 | 1 | = | ______ |
| E. | 1 | 0 | 0 | 1 | 0 | 0 | = | ______ |
| F. | 1 | 1 | 1 | 0 | 1 | 1 | = | ______ |

II. Now write numbers in binary code (base two) to represent the decimal numbers from 1 to 26.

1= ______________

2= ______________

3= ______________

4= ______________

5= ______________

6= ______________

7= ______________

8= ______________

9= ______________

10= ______________

11= ______________

12= ______________

13= ______________

14= ______________

15= ______________

16= ______________

17= ______________

18= ______________

19= ______________

20= ______________

21= ______________

22= ______________

23= ______________

24= ______________

25= ______________

26= ______________

III. First, double-check your work in part II. Then pretend that the binary code numbers above can be used to represent letters. Make 1 = A and 10 (in base 2) = B and so on so that the final number above in binary code equals Z. Do this by writing each letter after the proper binary number. Now, substitute those letters for the binary numbers below to "decode" these words:

A. 111 10101 1101

______________________________

B. 11001 101 10011

______________________________

C. 10110 1111 10100 101

______________________________

D. 10111 1 10100 101 10010

______________________________

E. 1100 1001 11010 1 10010 100

______________________________

F. 10011 10001 10101 1 10011 1000

______________________________

Name ______________________________

# Name That Document!

It is common to write and store dozens of letters and documents in a single word processing program. To be able to save and identify them for future use, a name must be given to each. In most cases, the document name must be no more than eight characters long, using only letters and numbers and no blanks. Because of these restrictions, the document names often look a little strange. For example, the name AMY1214 might refer to a letter written to Amy on December 14.

A. For each name below, write a possible description of the actual document. Feel free to use your imagination!

1. MILKEGGS ______________________________
2. SNOW123 ______________________________
3. RESUMEA ______________________________
4. VACATLOG ______________________________
5. BOOKPROP ______________________________
6. JSMITH45 ______________________________
7. CARCOMP ______________________________
8. IT ______________________________
9. MAXIMUM ______________________________

B. Now reverse the process. Write an acceptable name for each of these documents. Remember to use no more than 8 characters, to use numbers and letters only, and to leave no blank spaces.

10. personal letter to your brother, Gary, on March 29 ______________
11. important facts to consider when buying a dishwasher ______________
12. inventory of video games ______________
13. job application for Sam's Supermarket ______________
14. job application for Greg's Grocery ______________
15. checker tournament rules ______________

Name ______________________________

# Field Finesse

Databases are used to store and manipulate information. This information is stored in fields which can be sorted, arranged, alphabetized, etc. For example, a piano tuner might want to keep records of all his clients in such a way that shows him which customers need tunings this month and how to contact them. Here is what a template (pattern) might look like for client records:

| TITLE | FNAME | LNAME |
|---|---|---|
| ADDRESS | | |
| CITY | ST | ZIP |
| PHONE | HOME | WORK |
| LAST TUNING DATE | NEXT TUNING DATE | |
| FREQUENCY OF TUNING | BRAND OF PIANO | |

Each box represents a field. The computer can be told to sort or call up information for any field that exists. The tuner can ask the computer to make a list of all customers whose pianos were tuned in the month of May. He can ask for a list of all customers who live in the city of Hartford. He can ask for a list of all his customers, alphabetized by last name.

There are dozens of both personal and business uses for databases. Listed below are just a few. For each one, list at least five different fields that should be included in the database so that information can be called up or sorted in a helpful way.

1. inventory of CDs ______________________________
2. recipe collection ______________________________
3. Christmas card mailing list ______________________________
4. planting and gardening record ______________________________
5. hotel reservation list ______________________________
6. exercise log ______________________________
7. family birthdays and anniversaries ______________________________
8. inventory of stocks, bonds, and insurance policies ______________________________

Name ______________________________

# Pumpkin Program Problems

A computer program lists instructions telling the computer what to do. Any errors in writing the program will lead to mistakes in the computer's work. Here is a simple list of instructions written as if it is a computer program for a robot to follow:

1. Leave home.
2. Go to a pumpkin patch.
3. Ask for a ten-inch pumpkin.
4. If patch has none, go back to #2.
5. Go home.

A. This program has some basic flaws—problems that will not make the robot do what its owner probably intends for it to do. What are two "bugs" or mistakes?

1. ______________________________
2. ______________________________

B. Rewrite the entire program to correct the errors you have found.

Line 1 ______________________________
Line 2 ______________________________
Line 3 ______________________________
Line 4 ______________________________
Line 5 ______________________________
(Add more if necessary.)

Here is a flowchart that shows how the program works. It is most helpful to draw the flowchart before writing the program to prevent errors like the ones above.

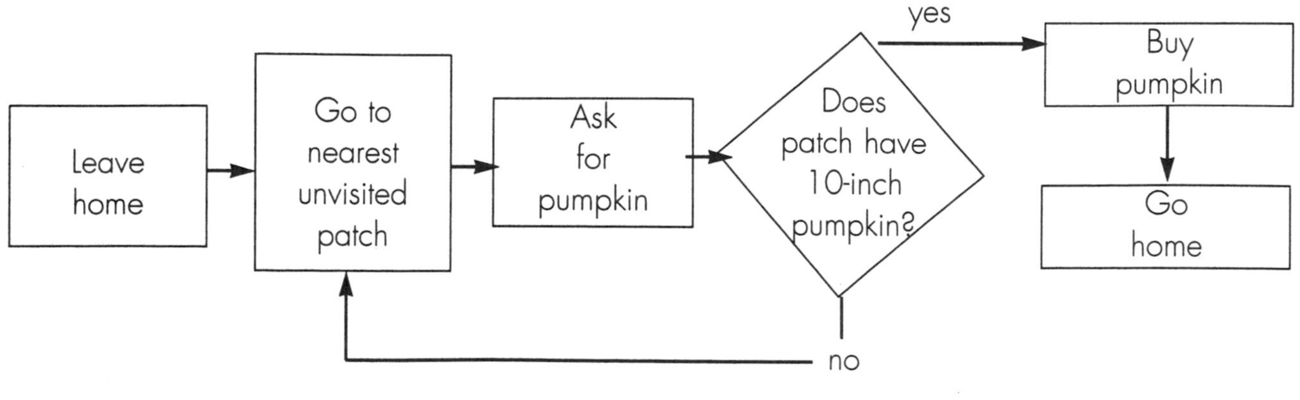

C. Now write another program that will tell the robot to visit bakeries and purchase at least 20 pumpkin pies. First, draw a flowchart, and then write a simple program. You may wish to number your steps in multiples of 10 (10, 20, 30, etc.) and leave spaces between steps in case you need to add extra lines later. This will save you renumbering and/or erasing.

Flowchart:

Program:

Name ______________________________

# Keyboard Quest

Knowing the computer keyboard is essential to quickly and accurately entering data. Here are some activities to help you learn the location of letters.

Now find words to complete this chart. Spell words using letters typed with just the right hand, the left hand, the home row keys, or those that alternate left and right (or right- and left-hand) keys. WARNING: It may not be possible to complete *all* the boxes. You are allowed to repeat letters within words. One word in each row has been filled in as an example.

| | 5 words of 4 letters | 3 words of 5 letters | 2 words of 6 or more letters | longest word possible |
|---|---|---|---|---|
| A. Left-hand keys | beat | | | |
| B. Right-hand keys | hill | | | |
| C. Home-Row keys | hall | | | |
| D. Alternating keys | with | | | |

# 3 Space and Space Exploration

This popular topic will encourage student interest in the planets, the solar system, and the history of space travel. As in previous sections, keep a good supply of library books and reference material available for study. Before beginning this section, spend some time on your own reviewing features of planets and important events in space travel. Almanacs and books written for the middle grades can present a lot of basic information in a condensed form.

Warm-up activities:

A. Review the planets by discussing these items with your class.
   1. Name the planets in order from the closest to the farthest from the sun. Now think of a saying using the first initials to help you remember their order. (For example, My Very Easy Math Jigsaw Simply Uses No Pieces.)
   2. What planet will have the longest year? Why? [Pluto, because it has the longest distance to travel completely around the sun.]
   3. Which planet is smallest? [Pluto] largest? [Jupiter] hottest? [Mercury] coldest? [Pluto] has the most rings? [Saturn] the most moons? [Saturn]

B. Think about space travel by discussing these questions with the class:
   1. Why are people interested in exploring space?
   2. What benefits are there in space exploration?
   3. What problems are there with space travel?
   4. Name all the astronauts you can remember.
   5. Name any U.S. space projects that you know about. [Mercury, Gemini, Apollo, etc.]
   6. Which two countries were rivals in the early days of space travel? [USA and USSR]
   7. Would you say that either country "won"? Why or why not?

### *Suggestions for Specific Puzzlers*

**Planetary Numbers I**
This can be completed by individuals or pairs. If desired, use it as a class project, recording answers on a large chart to post throughout the unit. Whichever method is chosen, it should be interesting to note how much information the students already know without consulting reference material. For columns C and D, ask students to speculate on the answers before verifying them with outside resources.

**Planetary Numbers II**
This will send students digging into your classroom reference materials. Do not ignore the hint; it is a fun way to keep students motivated, to keep them looking for complete answers, and to make certain that they double-check their work.

**Astronomical Alterations**
This may be simplest to complete by first deciding on the correct term for each definition and then choosing the word that can be transformed into that term. The beginning words, in most cases, look nothing like their solutions!

**Solar Calculations**
At the outset, decide whether or not calculators will be allowed. Try some problems on your own (2E and 4D, for example) to determine if your classroom calculators will accept enough digits. Answers here will vary, depending on how exact the numbers are that are used. For example, in number one, the distance of 93,000,000 miles is used here. For 4C and 4D, that distance has been multiplied by 1.6, a simple and standard conversion factor. However, both the 93 million and the 1.6 could be replaced by more exact numbers which will, of course, result in slightly different answers.

**Space-Age Phenomena**
Go over the example carefully with the class to be certain the directions are clear. Some students will be able to solve these quite quickly.

**Constellation Conundrum**
A reference book may be needed here, but see how far students can get without one. As a follow-up activity, see how many other constellations students can name for a composite classroom list.

**Travel Sequence**
Two sets of blanks are provided at the beginning of each line—one for speculation and one for actual answers. Be certain students fill in as many of the correct years as possible on the blanks that follow each event.

**Space Search**
Allow ample time for the completion of this activity. It is a fun way to cover a lot of important facts in space travel. You may wish to follow up on this with a quiz over the fill-in facts.

**Space History Mysteries**
This gives a good look into the backgrounds of important names in space history. The dates also help put these lives into historical perspective. Additional student activities could include calculating the person's age at time of flight, death, etc; relating events in a life with other important world events; writing mystery histories for others in this field.

**The Perfect Astronaut** and **In Your Own Words**
These open-ended activities are designed to make students think critically about career preparation, qualifications, and experiences. Allow students to share outcomes as desired.

Name ______________________________

# Planetary Numbers 1

Complete the number chart by placing the digits 1 to 9 in the correct spaces in each of the first three columns. In column D, you will need other numbers. Round to the nearest day or year. You will see that in column E the numbers are filled in, but the description is missing. Can you figure out what the numbers in this column represent? Write a suitable title at the top of column E.

| Planet | A. In ABC order | B. In order from the sun | C. In order from smallest to largest | D. Length of one year in earth time | E. |
|---|---|---|---|---|---|
| Mars | | | | | 2 |
| Earth | | | | | 1 |
| Pluto | | | | | 1 |
| Venus | | | | | 0 |
| Saturn | | | | | 18+ |
| Mercury | | | | | 0 |
| Uranus | | | | | 15 |
| Neptune | | | | | 8 |
| Jupiter | | | | | 16 |

Name ____________________

# Planetary Numbers II

Each planet in our solar system has been assigned a number as shown. Below are several descriptions of planets. Write the number(s) of the planet(s) that fit each description. For example, if the description read, "Third planet from the sun," the answer would be "3" for Earth. If it said, "Planet that revolves around our sun," the answers would be 1, 2, 3, 4, 5, 6, 7, 8 ,9, since all the planets are correct.

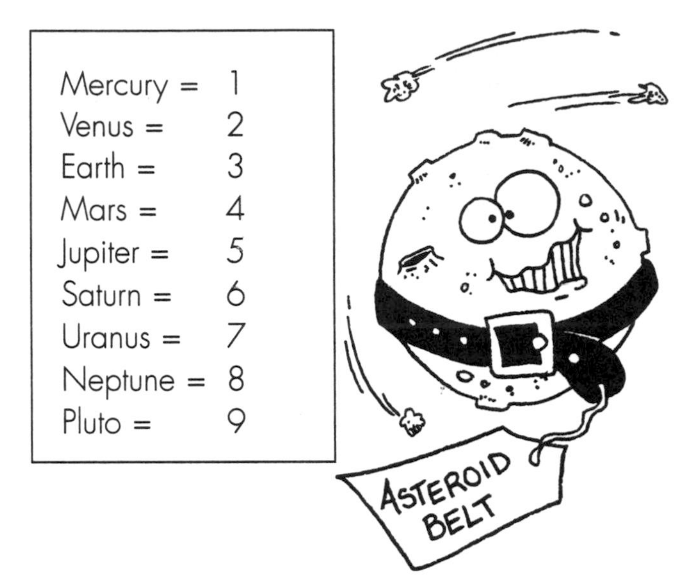

|  |  |
|---|---|
| Mercury = | 1 |
| Venus = | 2 |
| Earth = | 3 |
| Mars = | 4 |
| Jupiter = | 5 |
| Saturn = | 6 |
| Uranus = | 7 |
| Neptune = | 8 |
| Pluto = | 9 |



1. Diameter of planet is less than 10,000 miles ____________________
2. Planet with many natural satellites, the four largest of which are named the Galilean satellites after their discoverer ____________________
3. Asteroid belt is between these two planets ____________________
4. Planet explored by *Voyager 1* and/or *Voyager 2* ____________________
5. The brightest of all planets as observed from Earth____________________
6. Planet with a day nearly the same length as Earth's ____________________
7. Planet known to have one or more rings ____________________
8. Home of "The Great Red Spot" ____________________
9. Planet known to sustain life ____________________
10. Planet known as a "giant planet" ____________________
11. Planet was originally named Georgium after King George III ____________________
12. Planet known to have volcanoes on its surface ____________________

Hint: Go back over your answers above and add all the numbers together that you have written. The total should be 141. How close are you? It is possible to reach the correct total without having all the right answers, but this will give some idea of the accuracy of your work.

Name ______________________________

# Astronomical Alterations

Several astronomical definitions appear below. The terms for each also appear below, but each has been altered. Your job is to properly change each term and match it to the correct definition. To do this, change just one letter in each word and then rearrange the letters. The first one has been solved as an example.

| | New word | Definition | |
|---|---|---|---|
| 1. heard | earth | K | A. any body that revolves around the sun |
| 2. square | ____________ | ___ | B. a planet's natural satellite |
| 3. slurps | ____________ | ___ | C. a "shooting star" |
| 4. riots | ____________ | ___ | D. an extremely bright, compact object far beyond our galaxy |
| 5. mask | ____________ | ___ | E. the star nearest Earth |
| 6. pleats | ____________ | ___ | F. collapsed neutron star that emits pulsing radio waves |
| 7. loom | ____________ | ___ | G. a heavenly body with a starlike head and often with a long, luminous tail |
| 8. ants | ____________ | ___ | H. the path of a heavenly body revolving around another |
| 9. bus | ____________ | ___ | I. the imaginary line around which a body rotates |
| 10. metro | ____________ | ___ | J. a sphere of matter held together by its own gravitational field and generating nuclear fusion reactions in its interior |
| 11. taxi | ____________ | ___ | K. the third planet from the sun |
| 12. remove | ____________ | ___ | L. nicknamed the "red planet" |

Name ____________________

# Solar Calculations

On this page you will be working with some astronomically large numbers. Round your answers to the nearest whole number or place (such as to the nearest thousand or million).

1. Write down the distance from the Earth to the Sun in miles. ____________

2. Suppose you were able to drive a car at 55 mph to the sun. How long would it take you to get to the sun:

   A. in hours? ____________

   B. in days? ____________

   C. in years? ____________

   D. in minutes? ____________

   E. in seconds? ____________

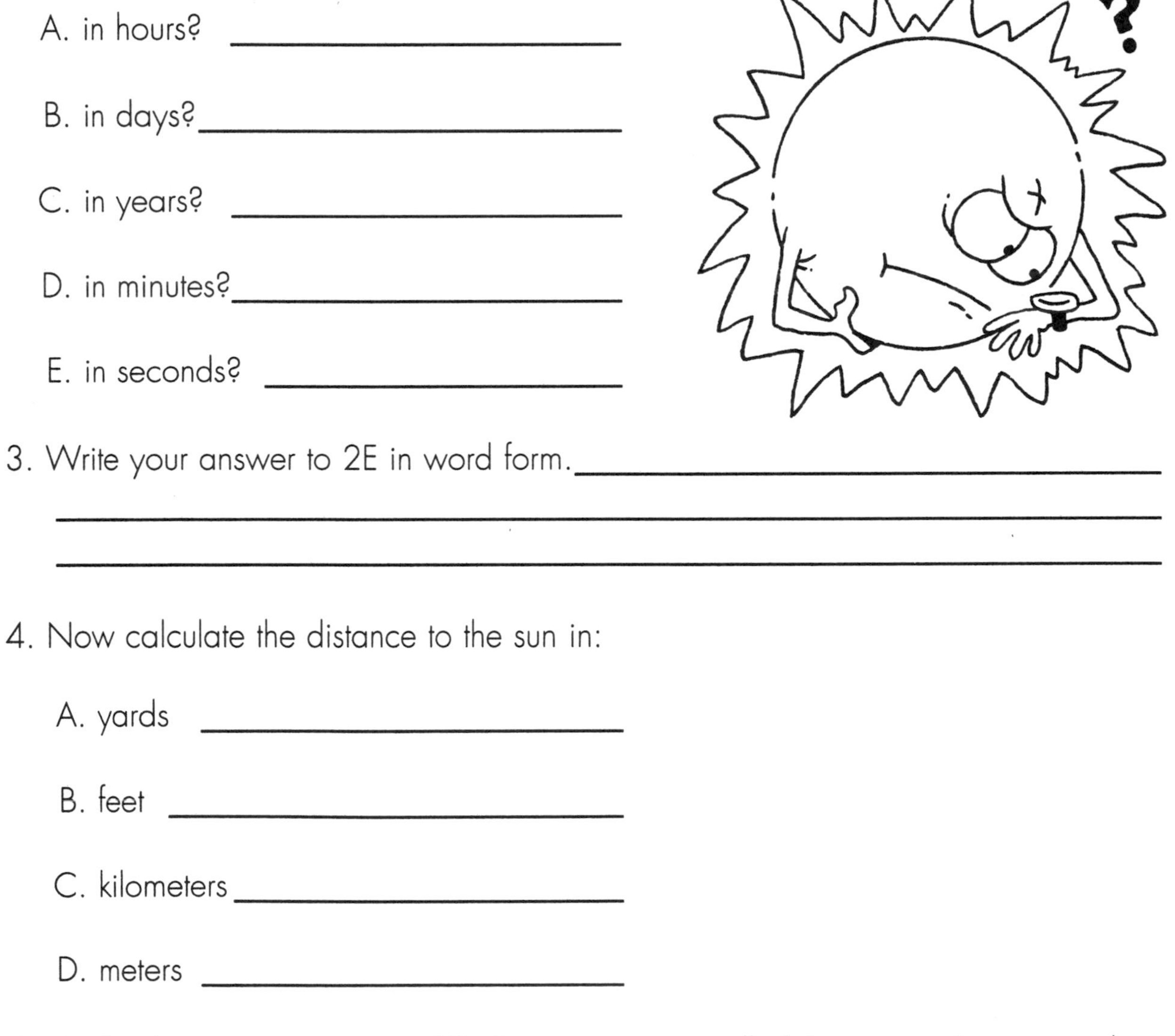

3. Write your answer to 2E in word form. ____________________

   ________________________________________

   ________________________________________

4. Now calculate the distance to the sun in:

   A. yards ____________

   B. feet ____________

   C. kilometers ____________

   D. meters ____________

5. Go back to your answer in 4C. Suppose you travelled this metric distance to the sun at a speed measured in km/h that is equal to the speed of 55 mph. Would it take more time, less time, or the same amount of time to reach the sun? ____________

Name ______________________________

# Space-Age Phenomena

Space is full of unusual objects, weather, and phenomena. Several are hidden below as interwoven compound words or two-word phrases. Can you solve each one? Hint: The two halves of words or phrases have been placed together without changing the order of any of the letters. You may need to add an apostrophe.

Example: H A C L O L E M Y E S T    Answer: HALLEY'S COMET

1. B H O L A L C E K ______________________________
2. S N U P O E V A R ______________________________
3. B R A O I W N ______________________________
4. G R I E A N D T ______________________________
5. S P S U O N T S ______________________________
6. N E S T U T R A O R N ______________________________
7. S E C O L L I P A R S E ______________________________
8. B A S E T E R O L I T D ______________________________
9. N O L R I T G H E H R T N S ______________________________
10. D W W A H I R F T E ______________________________
11. S A R T U I N R G S N S ______________________________
12. S H O S T O T A I R N G ______________________________
13. T S H U T O N D R E R M ______________________________
14. S M E H O T E W O E R R ______________________________
15. F L S A R O L E A R S ______________________________

Name ______________________________

# Constellation Conundrum

Many of the 88 recognized constellations have interesting names made of two words or a compound word. Below are all the correct parts of the names of 15 different constellations, but each part has been incorrectly paired with another part. Your job is to write the names of all 15 constellations correctly in the blanks, using each word shown here. Words are correctly positioned at either the beginning or the end of the constellation's name. Caution: Some parts appear to go with more than one name; however, there is one solution where all the parts can be paired correctly.

| | | | | | |
|---|---|---|---|---|---|
| Bird of | Dipper | Flying | Pump | Air | Tool |
| Big | Bearer | Water | Lion | Serpent | Dog |
| Southern | Dogs | Mariner's | Crown | Berenice's | Paradise |
| Northern | Hair | Great | Serpent | Hunting | Compass |
| Sea | Fly | Little | Fish | Sculptor's | Snake |

______________ ______________ ______________

______________ ______________ ______________

______________ ______________ ______________

______________ ______________ ______________

______________ ______________ ______________

Name ____________________________

# *Travel Sequence*

Listed below are 20 important events in the history of space travel. Can you place them in the proper order? In the left-hand blanks, number the events from 1-20. In the blank at the end of each item, write the year the event took place.

____ 1. First "space-walk" by Alexei Leonov ________

____ 2. *Vikings I* and *2*, from the USA, make controlled landings on Mars ________

____ 3. Launch of the Hubble Space Telescope ________

____ 4. Armstrong and Aldrin, in *Apollo 11*, land on the moon ________

____ 5. Launch of the first American satellite, *Explorer 1* ________

____ 6. First launch of America's Space Shuttle ________

____ 7. First liquid-fuel rocket launched by Goddard, USA ________

____ 8. First close-up pictures of Mercury by *Mariner 10*, USA________

____ 9. First successful unmanned probes to the Moon: *Lunas 1*, *2*, and *3*; USSR ____

____10. First manned flight around the moon, *Apollo 8*, USA ________

____11. Disaster with America's shuttle, *Challenger* ________

____12. V.2 rockets developed by Wernher von Braun and his German team ______

____13. Launch of *Skylab*, America's first space station________

____14. *Voyager 2* flew past Neptune ________

____15. First manned space flight, by Yuri Gagarin, USSR ________

____16. First pictures obtained from the surface of Venus, by USSR's *Venera 9* ______

____17. Launch of the first artificial satellite, *Sputnik I*, USSR ________

____18. First American orbital flight by John Glenn ________

____19. First controlled landing on the moon by *Luna 9*, USSR ________

____20. First good close-range views of Saturn, from USA's *Voyager 1*________

Name ______________________________

# Space Search

Here is a space-age challenge about space exploration—a word search without a word list. Clues to the words you need to find are listed on the next two pages. Work back and forth between the clues and the puzzle. Circle the words you find in the puzzle. They may appear vertically, horizontally, or diagonally both backwards and forwards. Write each word you find in the appropriate blank on the following pages. Try to complete all the blanks.

| | | | | | | | | | | | | | | |
|---|---|---|---|---|---|---|---|---|---|---|---|---|---|---|
| R | E | T | S | O | O | B | M | A | S | T | U | Y | L | A | S |
| E | L | K | Y | E | X | P | L | O | R | E | R | I | N | P | C |
| T | L | R | E | N | I | R | A | M | B | O | S | K | L | O | L |
| I | E | N | I | R | D | L | A | C | V | O | O | A | L | E | O |
| B | B | K | W | H | I | T | E | E | Y | T | S | U | O | B | R |
| R | Y | S | C | A | Z | D | R | U | S | H | M | N | T | E | T |
| O | T | Y | P | O | E | Y | Z | O | D | B | O | E | Z | A | N |
| R | R | T | R | U | R | D | V | O | I | V | L | B | A | N | O |
| A | E | I | E | S | T | E | W | A | R | S | A | S | K | I | C |
| N | B | V | G | A | O | N | G | E | T | L | Y | U | I | R | N |
| U | I | A | A | X | P | N | I | A | Y | D | E | N | A | A | O |
| L | L | R | Y | I | N | E | R | K | T | N | N | E | L | G | I |
| A | R | G | O | P | A | K | S | E | I | S | T | V | R | A | S |
| F | E | M | V | O | N | F | Y | X | L | M | I | I | R | G | S |
| N | E | V | E | S | R | N | R | O | A | B | S | T | B | I | I |
| D | D | R | A | P | E | H | S | R | U | S | B | X | L | R | M |
| D | A | R | N | O | C | O | S | M | O | N | A | U | T | U | O |
| A | R | E | N | E | V | J | U | M | D | O | G | Z | H | Y | M |

CLUES: (Unless otherwise stated, give only the last names of people.)

1. To reach outer space, rockets must escape this force. ______________
2. To burn fuel, a rocket needs ___________which comes from an _____________.
3. A rocket that has two or more stages (parts) is called a __________ __________.
4. A rocket engine that provides the initial thrust for a launch and is then discarded is a __________________.
5. Landing a spacecraft in the seas was called a ______________________.
6. The NASA crew that controls spaceflights is called ___________ ___________.
7. The first artificial satellite was launched by ____________ (country) and was called ______________.
8. The first American satellite was the ______________________.
9. The first television communications satellite launched in 1965 was called ______________.
10. The first space traveler's name was ______________. She was a ____________.
11. The first human to travel in outer space was (first and last names) ___________ ____________ who rode in the Soviet spacecraft named _______________.
12. The first American in a manned spacecraft was ________________, but he did not circle, or _________________ the earth.
13. The second such astronaut was ______________ who rode in a spacecraft called the _______________ _______________.
14. The first American to orbit the earth was ____________________.
15. The first Russian to walk in space was _____________________.
16. The first American to walk in space was ____________________.

17. A U.S. space probe sent to study the moon is known as a ________________ ________________.

18. The U.S. President whose goal it was to put a man on the moon was (first name, initial, last name) ________________________________________.

19. The country to put the first man on the moon was ____________________.

20. The second man to walk on the moon was ____________________.

21. Four other *Apollo* astronauts were __________, __________, __________, and ______________.

22. A moon ______________ was used to travel about on the surface of the moon.

23. In 1975, an *Apollo* vessel docked with a Soviet spacecraft called __________.

24. The names of two Russian space stations are __________ and __________.

25. The term used for a Russian astronaut is ____________________.

26. The first American space station was called ______________________.

27. The name of America's first space shuttle was ____________________. It could carry (how many?) __________ people.

28. The names of two U.S. space probes were __________ and __________.

29. The names of two USSR space probes were __________ and __________.

30. The U.S. spacecraft *Magellan* mapped most of this planet. ______________

31. The name of the U.S. space telescope launched in 1990 is ______________.

Name ______________________

# Space History Mysteries

Each box below contains some significant dates in the life of an important person in the history of space exploration. Simply discover the person's name (first and last, if possible) and write it on the blank inside the box.

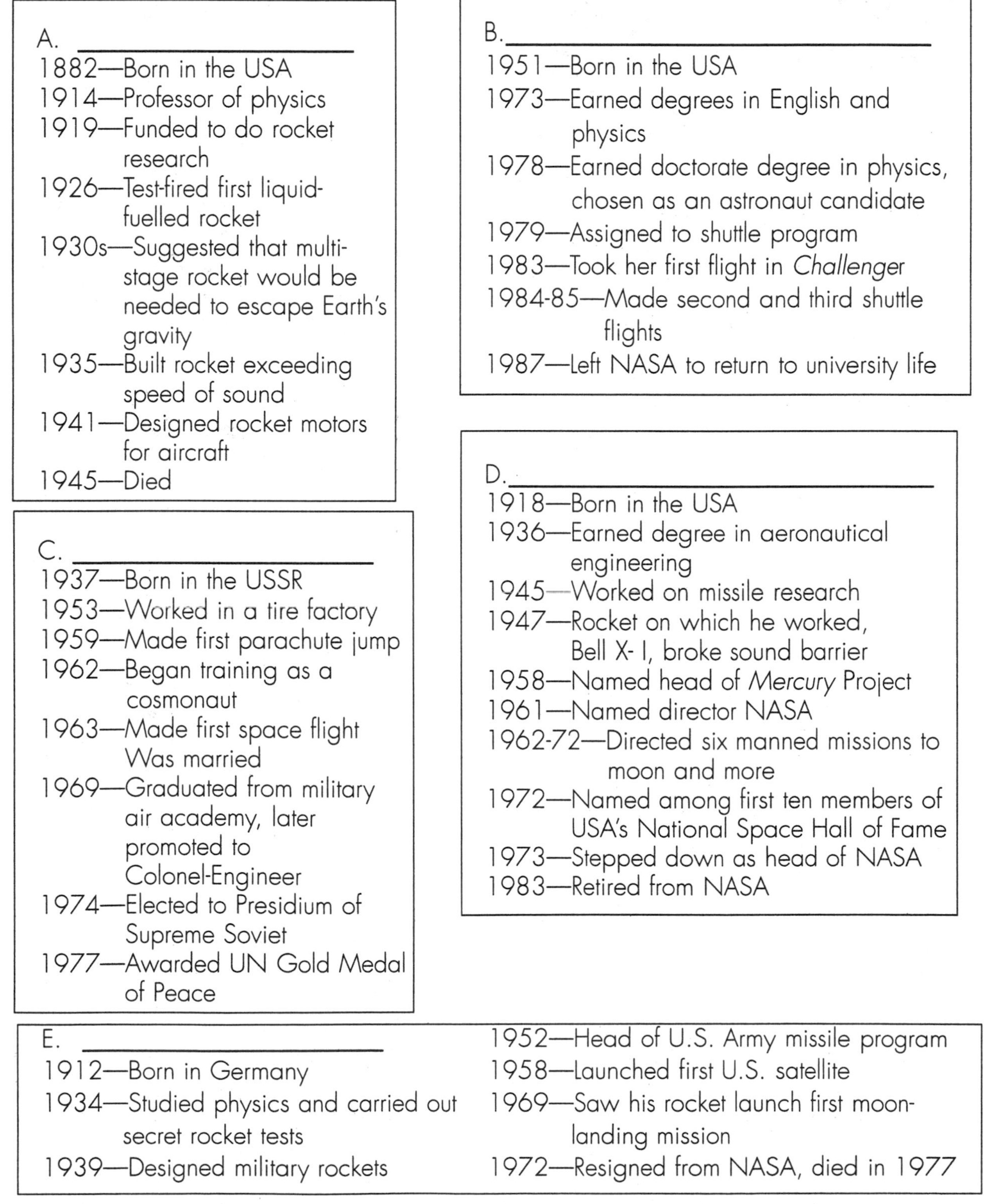

A. ______________

1882—Born in the USA
1914—Professor of physics
1919—Funded to do rocket research
1926—Test-fired first liquid-fuelled rocket
1930s—Suggested that multi-stage rocket would be needed to escape Earth's gravity
1935—Built rocket exceeding speed of sound
1941—Designed rocket motors for aircraft
1945—Died

B. ______________

1951—Born in the USA
1973—Earned degrees in English and physics
1978—Earned doctorate degree in physics, chosen as an astronaut candidate
1979—Assigned to shuttle program
1983—Took her first flight in *Challenger*
1984-85—Made second and third shuttle flights
1987—Left NASA to return to university life

C. ______________

1937—Born in the USSR
1953—Worked in a tire factory
1959—Made first parachute jump
1962—Began training as a cosmonaut
1963—Made first space flight Was married
1969—Graduated from military air academy, later promoted to Colonel-Engineer
1974—Elected to Presidium of Supreme Soviet
1977—Awarded UN Gold Medal of Peace

D. ______________

1918—Born in the USA
1936—Earned degree in aeronautical engineering
1945—Worked on missile research
1947—Rocket on which he worked, Bell X- l, broke sound barrier
1958—Named head of *Mercury* Project
1961—Named director NASA
1962-72—Directed six manned missions to moon and more
1972—Named among first ten members of USA's National Space Hall of Fame
1973—Stepped down as head of NASA
1983—Retired from NASA

E. ______________

1912—Born in Germany
1934—Studied physics and carried out secret rocket tests
1939—Designed military rockets
1952—Head of U.S. Army missile program
1958—Launched first U.S. satellite
1969—Saw his rocket launch first moon-landing mission
1972—Resigned from NASA, died in 1977

Name ______________________________

# The Perfect Astronaut

Before men and women fly in space, they spend years in training and are very thoroughly interviewed to make certain they are well-equipped for the challenges and dangers of space travel. Think carefully about what makes a person well-suited for this unusual profession. Is it their knowledge, training, certain personality traits, or some combination of these?

A. In your opinion, what are the top ten characteristics/qualifications of the perfect astronaut?

1.
2.
3.
4.
5.
6.
7.
8.
9.
10.

B. If you were interviewing prospective candidates, what ten questions would you most like to ask them?

1.
2.
3.
4.
5.
6.
7.
8.
9.
10.

Name ______________________________

# *In Your Own Words*

Imagine that you are one of the very first astronauts, encountering many unusual and difficult circumstances. First, do some reading on living conditions in space, such as dealing with weightlessness, living in tight quarters, etc. Then write a first-person account of your own fictional experience on any two of the following topics. Write three or four paragraphs for each one.

1. leaving the launchpad
2. eating your first meal in outer space
3. moving about inside the space capsule
4. communicating with other crew members
5. taking a shower
6. changing your clothes
7. writing a letter
8. seeing the earth from space
9. catching your first close-up glimpse of the moon
10. walking outside the spacecraft
11. exploring the moon in a moon rover
12. docking with a foreign space vessel
13. performing scientific experiments in space
14. landing back on earth
15. deciding whether or not to make another space flight

# 4 Weather and Climate

This unit will help your students understand the basic factors that determine weather conditions around the world. They will also learn weather-related vocabulary and symbols. Fun facts are also included about unusual occurrences around the world, and students will then be given a chance to speculate about the possibility of being able to predict and control weather. Make certain they have access to appropriate reference materials such as almanacs and encyclopedias.

Warm-up activities:

A. Discuss these questions with the class.
   What are the three factors that determine our weather? (sun, water, and wind)
   How do "weather" and "climate" differ?
   Why does the climate at the poles and the equator differ so much from ours?

B. Review the water cycle. Allow students to help make a drawing on the chalkboard of water's continuous movement. Include water evaporation from rivers, plants, seas, etc., into the air where it forms clouds. Show the clouds losing that water vapor in the form of rain or snow which goes back to the earth, into the ground water, or into rivers, seas, etc., to start cycling again.

C. Bring in forecasts from newspapers or take notes on those on television and radio. Watch weather forecasts for a week or more. Record their accuracy.

## *Suggestions for Specific Puzzlers*

### Climate Conundrum

This is a good introductory activity. Students will, of course, need to work back and forth between the article and the scrambled words. If they get "stuck," you may wish to reveal either the first letter of the correct word or the number of the blank in which a scrambled word belongs. When finished with the activity, be certain that students read it again for understanding and perhaps keep it throughout the unit.

### Meteorologist's Mix-Up

Weather forecasts brought into class will be useful here in suggesting possible answers.

### Search High and Low

This is an excellent activity for developing skills in reading charts and making only correct inferences. Encourage students to work slowly and thoughtfully. The M answer simply means that the statement may or may not be true based on the facts presented, not on any outside knowledge or guesses that students may have. Remind students that scientists must deal with hard facts when making statements.

**Symbol Sense**

Reference books and weather forecasts printed in the newspaper may be necessary.

**Where in the World?**

Help students see how geography influences climate and weather. While it may not be possible to get all the right answers simply by studying a map, students should be able to narrow the possibilities for many of the answers. Note: All data here is from the 1996 *Guinness Book of World Records.*

**Chilly Challenger**

First, be certain that students know how to read the chart correctly. Do several questions together, such as: What is the wind chill factor when the temperature is 0°F and the wind speed is 30 mph? (-49°F) When the wind chill is -15°F and the temperature is 20° F, what is the wind speed? (25 mph)

**Temperature Tantrum**

Be certain that students have correctly completed the previous activity before beginning this one, since the completed wind chill factor chart is required here. As stated in the directions, questions become increasingly more difficult, requiring the use of more charts and conversions. If necessary, do a sample one together that is similar to those near the bottom of the page. For example: The temperature is -12°C. The wind chill factor is -33°F. What is the name of the wind? To solve this, first convert the Celsius temperature to Fahrenheit. The answer is about 10°F. Next, look up this temperature on the wind chill chart. It shows that for the wind chill factor to be -33°F when the temperature is 10°F, the wind speed must be about 30 mph. Now refer to the Beaufort Scale to see that a 30 mph wind is a strong breeze.

**Weather Trivia**

This includes an assortment of interesting facts and terms which can be found in encyclopedias, almanacs, and other reference books. This could be completed in small groups.

**Weather Wizard I and II**

These are open-ended activities designed to get students to think about the implications of accurately predicting and controlling the weather. Feel free to expand upon these in any appropriate manner, and allow students to share their outcomes if desired.

Name ______________________________

# Climate Conundrum

Here are a few paragraphs of basic information about climate and weather. As you can see, 30 words are missing from the article. All the words appear on the next page, but they are scrambled. First unscramble each word and write it correctly in the first blank of each line on the next page. In the second blank, write the number of the blank from the article in which it belongs. One has been done for you as an example.

__1__ is the condition of the air above a particular place on __2__ at a particular time. It __3__ from place to place and from time to time. __4__ is the weather __5__ at a particular place over a long __6__ of time.

Weather is a product of __7__ (8 letters), __8__ (3 letters), and __9__ (5 letters). One heats the earth, one wets the earth, and the other moves the heat and water around. The amount of heat and __10__ each place receives depends on the __11__ of the sun's rays and how long the sun __12__. The __13__ becomes very hot because the sun's rays hit it __14__. The __15__ are always cold because, even when the sun's rays strike them, they strike at an angle.

The __16__ of air is caused mainly by heat. When air becomes warmer it __17__, the molecules move farther apart, the air becomes lighter, and the air __18__ falls. When air becomes cooler, the __19__ move closer together, the air becomes __20__, and the air pressure __21__. When warm air rises from a place, cool air moves in. This movement of warm and cool air is __22__.

Three-__23__ of the earth's __24__ is covered by water, which is in constant __25__. Water from the oceans __26__ to form water __27__ in the air. This collects to form __28__, which are __29__ around by the wind until they drop their water back onto the land or into the __30__.

| | | |
|---|---|---|
| A. ira | air | 8 |
| B. daxpens | | |
| C. iverahe | | |
| D. timoon | | |
| E. thawere | | |
| F. yetclird | | |
| G. talemic | | |
| H. acone | | |
| I. dinw | | |
| J. ovpar | | |
| K. serserup | | |
| L. dropie | | |
| M. teraw | | |
| N. heart | | |
| O. slope | | |
| P. siser | | |
| Q. wobnl | | |
| R. facruse | | |
| S. ravies | | |
| T. orqueta | | |
| U. thigl | | |
| V. glean | | |
| W. teenvmom | | |
| X. cellosume | | |
| Y. tarsquer | | |
| Z. atrospavee | | |
| AA. dulsoc | | |
| BB. taprent | | |
| CC. inhess | | |
| DD. hultsing | | |

Name ______________________________

# Meteorologist's Mix-Up

World-famous meteorologist, I.M. Inafogg, just hit the wrong key on the word processor while preparing his next weather report. Several eight-letter words or phrases have been separated into four pairs of letters and mixed up on the list shown here. Can you straighten out this mess? Find the four pairs of letters that go together, and then write the completed word in the blank by its beginning letters. An example has been completed for you.

| | | | | | |
|---|---|---|---|---|---|
| latitude | Ex. | LA | NS | FA | RM |
| ________ | 1. | FO | MI | SO | NG |
| ________ | 2. | SU | ES | WI | DE |
| ________ | 3. | RA | OW | ZA | ND |
| ________ | 4. | IC | TI | SU | RD |
| ________ | 5. | PR | NS | CA | LL |
| ________ | 6. | DE | EE | TO | NE |
| ________ | 7. | BL | RE | DI | LL |
| ________ | 8. | HU | GH | FA | RY |
| ________ | 9. | AD | ES | TU | TY |
| ________ | 10. | FR | VI | EF | RE |
| ________ | 11. | HI | IZ | ZI | ST |
| ________ | 12. | SN | IN | HI | OG |

Name ______________________________

# Search High and Low

This table shows some average high and low monthly temperatures for ten selected cities in the United States over a recent 30-year period. Search through these figures to find out the facts. The statements that are listed below the table are either true, false, or cannot be ascertained either way from the information given. If the statement is definitely true, write a T in the blank beside it. If the statement is definitely false, write an F in the blank. If the statement may or may not be true, write an M in the blank.

Average monthly temperature in degrees Fahrenheit

| City | Jan. | April | July | Oct. |
|---|---|---|---|---|
| Atlanta, Georgia | 42 | 62 | 79 | 62 |
| Austin, Texas | 49 | 69 | 85 | 70 |
| Chicago, Illinois | 22 | 49 | 73 | 54 |
| Fairbanks, Alaska | -13 | 30 | 62 | 25 |
| Fargo, North Dakota | 4 | 42 | 71 | 46 |
| Honolulu, Hawaii | 73 | 76 | 80 | 80 |
| Miami, Florida | 67 | 75 | 83 | 78 |
| New York, New York | 32 | 52 | 76 | 58 |
| Portland, Maine | 22 | 43 | 68 | 49 |
| Richmond, Virginia | 37 | 58 | 79 | 59 |

____ 1. Of the cities shown, Fairbanks has the largest change in monthly temperature from January to July.

____ 2. Of the cities shown, Miami has the smallest change from January to July.

____ 3. The coldest U.S. city during the month of January is Fairbanks, Alaska.

____ 4. The greatest variation in temperature among the cities shown between July and October occurs in Fargo.

____ 5. During January, the temperature in New York is ten degrees warmer than in Portland.

____ 6. Of the cities shown, Honolulu has the most constant average temperatures.

____ 7. In Chicago, summers are 19 degrees warmer than autumns.

____ 8. The average temperature difference in Austin from January to April is 20 degrees.

____ 9. The average temperature in Portland during the month of July is 29 degrees warmer than the average temperature in October.

____ 10. The range of average temperatures in April for these ten cities is 26 degrees.

____ 11. The warmest U.S. city in July is Austin.

____ 12. The range of average temperatures in October for these ten cities is 55 degrees.

____ 13. The greatest range of average temperatures for these ten cities throughout the year occurs in the month of January.

____ 14. Of these ten cities from January to October, Chicago has the biggest change in average temperature.

____ 15. Three cities above all have average temperatures in January that are 20 degrees cooler than their April average temperatures.

Name ____________________________

# Symbol Sense

Forecasters use symbols to show others their weather predictions. These symbols are often used in television and newspaper forecasts. How well do you know these symbols?

The boxes below contain 12 standard weather symbols plus four "fake" symbols. At the bottom of the page are descriptions of the symbols, but there are two extra symbols. First cross out the fake symbols. Then match the real symbols with their correct labels by writing the number of each in a blank by its label. Finally, cross out the two extra symbols.

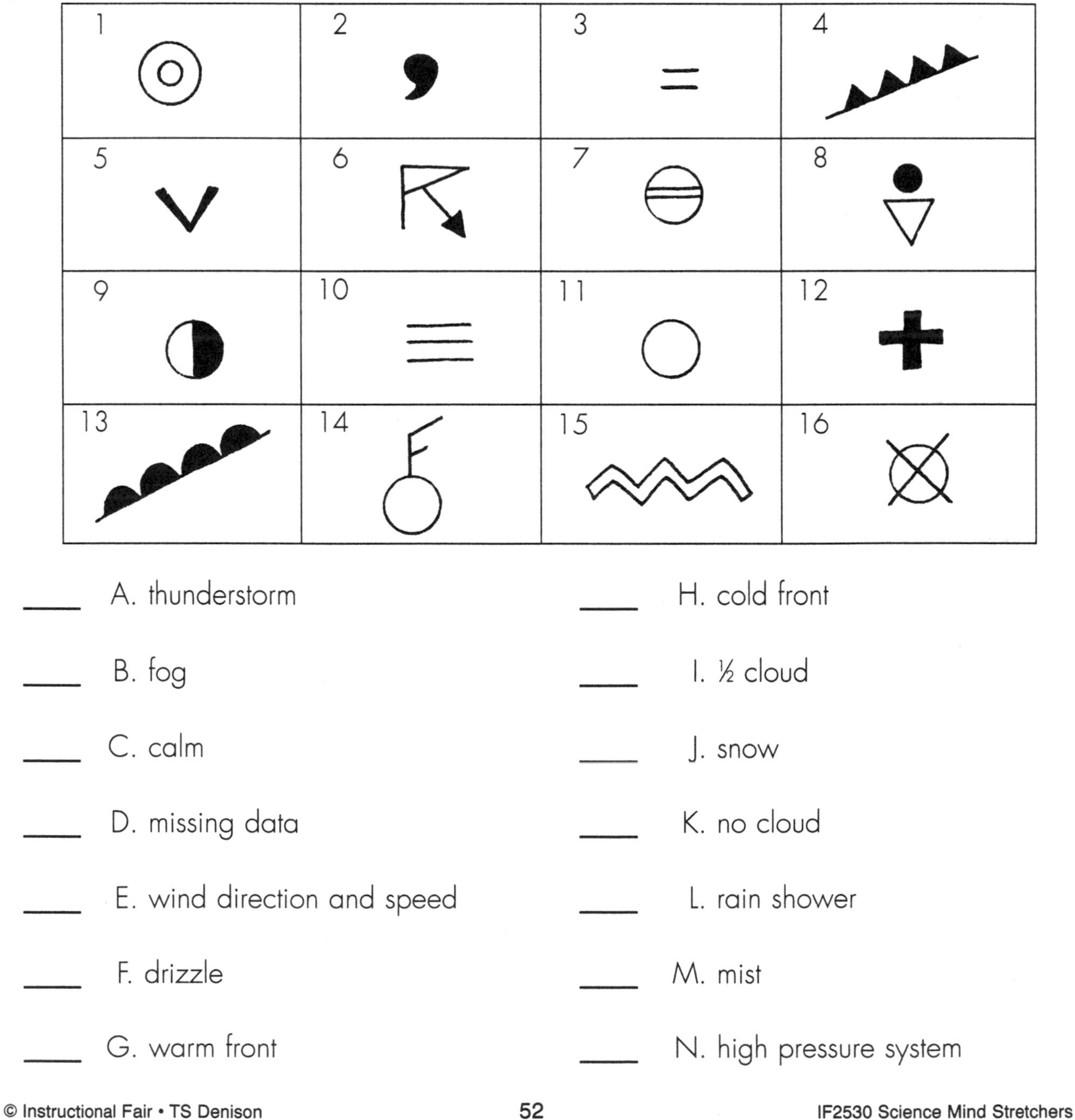


____ A. thunderstorm

____ B. fog

____ C. calm

____ D. missing data

____ E. wind direction and speed

____ F. drizzle

____ G. warm front

____ H. cold front

____ I. ½ cloud

____ J. snow

____ K. no cloud

____ L. rain shower

____ M. mist

____ N. high pressure system

Name ______________________________

# Where in the World?

Do you know where in the world you would find the hottest place, the coldest place, or the wettest place? Use a globe or map and some common sense to match the descriptions and locations below. Write the letter of the location in the blank next to the correct description.

| | Description | Location |
|---|---|---|
| ____ | 1. The wettest place in the world, with an average annual rainfall of 468 inches/yr. | A. Kauai, Hawaii |
| ____ | 2. The driest place in the world where the average annual rainfall is 0.004 in. | B. Verkhoyansk, Russia |
| ____ | 3. The world's sunniest place, with sunshine during 91 percent of the possible hours | C. Mawsynram, India |
| ____ | 4. The world's hottest place, with an annual average in one 6-year period of 94°F. | D. Yuma, Arizona |
| ____ | 5. The coldest location in the world, with an annual average of -72°F. | E. Tamarac, California |
| ____ | 6. The location of the greatest temperature ranges where temperatures have varied from -90°F to 98°F | F. Cherrapunji, India |
| ____ | 7. The place receiving the world's greatest annual rainfall of 1042 inches (1860-61) | G. Gopalganj, Bangladesh |
| ____ | 8. The spot with the most rainy days per year | H. Pacific Coast of Chile near Arica |
| ____ | 9. The location with the greatest snowfall during a 12-month period (1971-72) | I. Mt. Rainier, Washington |
| ____ | 10. The place with the greatest depth of snow ever measured on the ground (1911) | J. Dallol, Ethiopia |
| ____ | 11. The location with the highest average number of days of thunder per year | K. Tororo, Uganda |
| ____ | 12. The place where the heaviest hailstones in the world fell in 1986, killing 92 people | L. Pole of Inaccessibility, Antarctica |

Name ______________________________

# Chilly Challenger

Here is a chart of wind chill factors. The chart shows equivalent temperatures for different combinations of wind speed and temperature. For example, if the temperature is 10°F and the wind is blowing at 15 mph, the temperature feels like (and has a cooling power equal to) -18° F.

You will notice that some numbers are missing from the chart. Use the clues to complete the chart. Then save the completed chart to use with the next activity.

Wind Chill Factors

| Wind speed (in mph) | Thermometer reading in degrees Fahrenheit | | | | | | | | | | | | | |
|---|---|---|---|---|---|---|---|---|---|---|---|---|---|---|
| | 35 | 30 | 25 | 20 | 15 | 10 | 5 | 0 | -5 | -10 | -15 | -20 | -25 | -30 |
| 5 | □A | 27 | 21 | 19 | 12 | 7 | 0 | B□ | -10 | -15 | -21 | C□ | -31 | -36 |
| 10 | 22 | 16 | 10 | 3 | -3 | -9 | D□ | -22 | -27 | -34 | -40 | -46 | -52 | -58 |
| 15 | 16 | 9 | 2 | -5 | -11 | -18 | -25 | -31 | -38 | -45 | -51 | -58 | -65 | -72 |
| 20 | 12 | 4 | E□ | -10 | -17 | -24 | -31 | -39 | -46 | -53 | -60 | -67 | -74 | -81 |
| 25 | 8 | 1 | -7 | -15 | -22 | F□ | -36 | -44 | -51 | -59 | G□ | -74 | -81 | -88 |
| 30 | 6 | -2 | -10 | -18 | -25 | -33 | -41 | -49 | H□ | -64 | -71 | -79 | -86 | -93 |
| 35 | 4 | -4 | -12 | -20 | I□ | -35 | -43 | -52 | -58 | -67 | -74 | -82 | -89 | -97 |
| 40 | 3 | -5 | -13 | -21 | -29 | -37 | -45 | -53 | -60 | -69 | -76 | -84 | -92 | -100 |
| 45 | 2 | J□ | -14 | -22 | -30 | -38 | K□ | -54 | -62 | -70 | -78 | -85 | -93 | □L |

Clues:

1. The wind chill factor for a temperature of 5°F and a wind speed of 10 mph is the same as the wind chill factor for a temperature of -10°F and a 5 mph wind.
2. At a temperature of 10°F, increasing the wind speed from 20 mph to 25 mph lowers the wind chill factor by 5 degrees.
3. At a temperature of -5°F, decreasing the wind speed from 35 mph to 30 mph raises the wind chill factor by 2 degrees.
4. At a temperature of 35°F, the wind chill factor is only 2 degrees lower than the actual temperature when there is a 5 mph wind.
5. The wind chill factor at 25°F, with a 20 mph wind, is the same as the wind chill factor for a temperature of 15°F and a 10 mph wind.
6. At a temperature of 15°F, when the wind speed is increased from 20 mph to 35 mph, the wind chill factor is lowered by 10 degrees.
7. The lowest wind chill factor on the chart is missing; it is two degrees lower than the number directly above it.
8. When the temperature is -15°F and the wind speed is 25 mph, the wind chill factor is one degree lower than when the temperature is -25°F and the wind is at 15 mph.
9. The remaining missing numbers on the chart are exactly halfway between the numbers to the right and to the left of them.

Name ______________________________

# Temperature Tantrum

It has been a dull day at the local weather station so, to liven things up a bit, the director has issued his staff some challenging mind stretchers. He has presented them with 15 different questions, all of which can be answered by piecing together lots of information. It is necessary to use the wind chill chart from the previous page, the Beaufort scale shown here, and the Celsius-Fahrenheit conversion information. Become a top-notch weather detective yourself by correctly solving the questions on the next page. They start out easy but quickly become more difficult. See if you can find the temperatures and wind speeds without throwing a tantrum! (You will need to round some numbers.)

The Beaufort Scale (devised by Francis Beaufort in 1806)

| No. | Name of wind | Wind Speed | Effects of Wind |
|---|---|---|---|
| 0 | Calm | 0-1 mph | Smoke goes straight up |
| 1 | Light air | 1-3 mph | Smoke slightly bent |
| 2 | Light breeze | 4-7 mph | Leaves rustle |
| 3 | Gentle breeze | 8-12 mph | Leaves move |
| 4 | Moderate breeze | 13-18 mph | Small branches move |
| 5 | Fresh breeze | 19-24 mph | Small trees sway |
| 6 | Strong breeze | 25-31 mph | Large branches move |
| 7 | Moderate gale | 32-38 mph | Whole trees sway |
| 8 | Fresh gale | 39-46 mph | Twigs break off |
| 9 | Strong gale | 47-54 mph | Roofs damaged |
| 10 | Whole gale | 55-63 mph | Trees uprooted |
| 11 | Storm | 64-75 mph | Widespread damage |
| 12 | Hurricane | Over 75 mph | Wholesale destruction |

Temperature Conversion Information

$C^\circ = (F^\circ - 32) \times 5/9$

$F^\circ = (C^\circ \times 9/5) + 32$

Some common equivalents:

0°C = 32°F

10°C = 50°F

30°C = 86°F

40°C = 104°F

100° C = 212°F

1. The temperature is 10°C. What is the temperature in °F? ____________________

2. The temperature is 68°F. What is the temperature in °C? ____________________

3. The temperature is -12°C. The wind is blowing at a speed of 5 mph. What is the wind chill factor? ________________

4. The temperature is -15°C. The wind is blowing at a speed of 20 mph. What is the wind chill factor? ________________

5. The wind chill factor is -22°F. The wind speed is 25 mph. What is the temperature?

   ________________

6. The wind force is 5 on the Beaufort scale. The temperature is -20°F. What is the wind chill factor? ________________

7. The wind force is 8. The temperature is -9°C. What is the wind chill factor? ______

8. There is a gentle breeze. The temperature is 15°F. What is the wind chill factor?

   ______________

9. There is a light breeze. The temperature is -4°C. What is the wind chill factor?

   ______________

10. There is a strong breeze. The wind chill factor is -56°F. What is the temperature?

    ______________

11. Whole trees are swaying. The temperature is -7°C. What is the wind chill factor?

    ______________

12. The temperature is 5°F. The wind chill factor is -15°F. What number on the Beaufort scale is the wind force? ____________

13. The temperature is 25°F. The wind chill factor -3°F. What is the name of the wind?

    ______________

14. The temperature is -20°C. The wind chill factor is -56°F. What number of the Beaufort scale is the wind force?

    ______________

15. The temperature is -1°C. The wind chill factor is -4°F. What is the name of the wind? __________________________

Name ______________________________

# Weather Trivia

Be a weather expert and find answers to all these questions. You will need reference books to help you find weather and climate facts, figures, and trivia.

1. Find the names of the four layers of the atmosphere. List them in order from closest to farthest from the surface of the earth.
   A. ______________________________
   B. ______________________________
   C. ______________________________
   D. ______________________________
2. The earth's axis is tilted at an angle of ________ degrees.
3. At the time of the spring and fall equinoxes, there are 12 hours of daylight and 12 hours of darkness all over the world. The dates of these are ______________ and ______________.
4. *Equinox* means ______________________________.
5. What is an anemometer? ______________________________
6. What does a barometer measure? ______________________________
7. What does a hygrometer measure? ______________________________
8. Tornadoes are most common in what country? ______________________
   About how many are there in an average year? ______________________
9. Air pressure inside the funnel cloud of a tornado is extremely ____________.
10. Whirlwind systems that form in the Atlantic Ocean are called ____________.
11. When these form in the Far East they are called ____________ or ____________.
12. The three basic cloud forms are ________________, ________________, and ________________.
13. The most common form of precipitation is ______________________________.
14. In a high pressure weather system, the wind blows in a circle, moving in a ________________ direction.
15. In which regions of the earth can thunderstorms occur nearly every day?
    ______________________________________________
16. Which regions of the earth receive the heaviest snowfalls? ______________

Name ______________________________

# Weather Wizard I

Imagine that you have just made the greatest discovery of the century—how to predict the weather with 100 percent accuracy!

A. First, tell how you made the discovery and how it works.

_______________________________________________

_______________________________________________

_______________________________________________

_______________________________________________

_______________________________________________

B. Tell how you can convince skeptics that your method works.

_______________________________________________

_______________________________________________

_______________________________________________

_______________________________________________

_______________________________________________

C. You plan to get rich on your discovery. How will you market and protect your idea?

_______________________________________________

_______________________________________________

_______________________________________________

_______________________________________________

_______________________________________________

D. On the back, sketch an ad for your discovery that could be printed in magazines or newspapers.

_______________________________________________

_______________________________________________

_______________________________________________

_______________________________________________

_______________________________________________

Name ______________________________

# Weather Wizard II

Now suppose that you cannot only predict the weather with perfect accuracy, but that you can also control it.

A. If you alone had the ability to control the weather, would you want to? Why or why not?

______________________________

______________________________

______________________________

______________________________

______________________________

B. List five advantages of being able to control the weather.

1. ______________________________
2. ______________________________
3. ______________________________
4. ______________________________
5. ______________________________

C. List five disadvantages of being able to control the weather.

1. ______________________________
2. ______________________________
3. ______________________________
4. ______________________________
5. ______________________________

D. Investigate "cloud seeding"—a method by which people have "made it rain." When was the method first discovered? ______________________________
How has the method been used? ______________________________

______________________________

# 5 Animals and Plants

All living things are divided into two kingdoms—animals and plants. These are vast topics which cannot be fully covered in a few pages. Here, the focus is mainly on recognizing and categorizing names of various plants and animals. These pages are meant to be springboards into further research. Students will encounter many unfamiliar animals and plants; hopefully, they will be encouraged to learn more about them. When studying new organisms, have students learn about their features, their habitats, and their similarities to and differences from familiar organisms. In the animal pages, students will also learn about names of animal groups and the correct terms for the male, female, and young in several species.

Some good wildlife books with basic information and accurate pictures would be most beneficial while working on this section. When students learn there is such a thing as a death cup mushroom, for example, they will naturally be curious to see what one looks like and where it grows. Having resources already on hand allows you to capitalize on such teachable moments.

Warm-up activities:

A. Brainstorm as a class on the huge variety of living things. Call out a category and have students list as many plants or animals in the category as possible within a one- or two-minute time limit. Possible animal categories: mammals, insects, amphibians, endangered species, Australian animals, mollusks, etc. Possible plant categories: trees, ferns, mosses, tropical plants, etc.

B. Encourage students to go on a "scavenger hunt" through reference books, looking for the most unusual plant or animal. Have them write brief reports on them.

## *Suggestions for Specific Puzzlers*

### Think Animals!

This is a good brainstorming activity, and it can be surprisingly difficult. You may wish to allow students about 10-15 minutes to complete as much as they can on their own, count up their completed boxes, and then finish as much of the page as possible with the help of reference material. An interesting variation is to put students in groups of three to five students and give them a 15-minute time limit. Then have them "score" their page; award ten points for each completed box and deduct five for every blank. The group with the highest score wins.

### Group Therapy

This is a fun format in which to find many new words. You will need to have almanacs, encyclopedias, and/or dictionaries on hand for student use. Encourage students to work back and forth between the story and the list. For example, if they know that bees form a *swarm*, then they can fill in the blank and then look for the word. Or, if they notice the word *colony* hidden in the story, they can look it up in a dictionary to learn that it can refer to a group of ants.

**Label Liabilities**
Some answers will be very easy; others are trickier, especially since the wrong choices are often very similar to the right choices. Almanacs and encyclopedias are good resources.

**Mammal Scramble**
Once the mammals are correctly identified, the puzzling element requires some attention. Encourage students to make more puzzle boxes of their own, using a different category of animals or a different number of letters. As students scour through materials looking for names that "fit" together, they will undoubtedly encounter a lot of new animals.

**Places, Please**
Reference books will be a big help. While students can solve parts of the page simply by fitting animals into the only possible places, there are some that are interchangeable. Here again, encourage students to dig a little deeper and learn more about the animals that are new to them.

**Would You Rather Be a Bee?**
This is an open-ended activity in which you can expect a wide variety of outcomes. Encourage class discussion of similarities and differences. As a class, how many can they find? Allow students who are willing to share journal entries.

**Think Plants!**
(Follow the same hints as in Think Animals!)

**A to Z Plants**
Encourage the use of pencils and erasers. Most solutions will be names that are familiar to students. *Thuja* will probably be new. Ask students to write more puzzles like this.

**Plantegories**
Here students may know many of the answers on their own, but they will probably need outside resources to completely finish the page. Many discussions and activities can spring from this page. What other carnivorous plants are there? How do they "eat meat"? Do all evergreen trees bear cones? How do ferns reproduce?

**Fractured Trees**
This is truly a "toughie" to challenge your brightest students. Not only do they have to be familiar with names of trees, but they must also be able to solve the brief crossword-type clues. If necessary, give hints to one or two of the key words.

**Mushroom Mania**
Reference books are a requirement here! Use this to lead into a deeper study of mushrooms, other fungi, and spore reproduction.

**Top Ten**
Use this open-ended writing to spark classroom discussion on related topics such as ecology and conservation.

# Think Animals!

As quickly as possible, fill in as many boxes below as you can. For each letter, write the name of an insect, mammal, reptile, and bird that begins with that letter. For example, in the first box you may choose to write *ant*. It may not be possible to fill every box; if you can complete all but three or four, you have done extremely well.

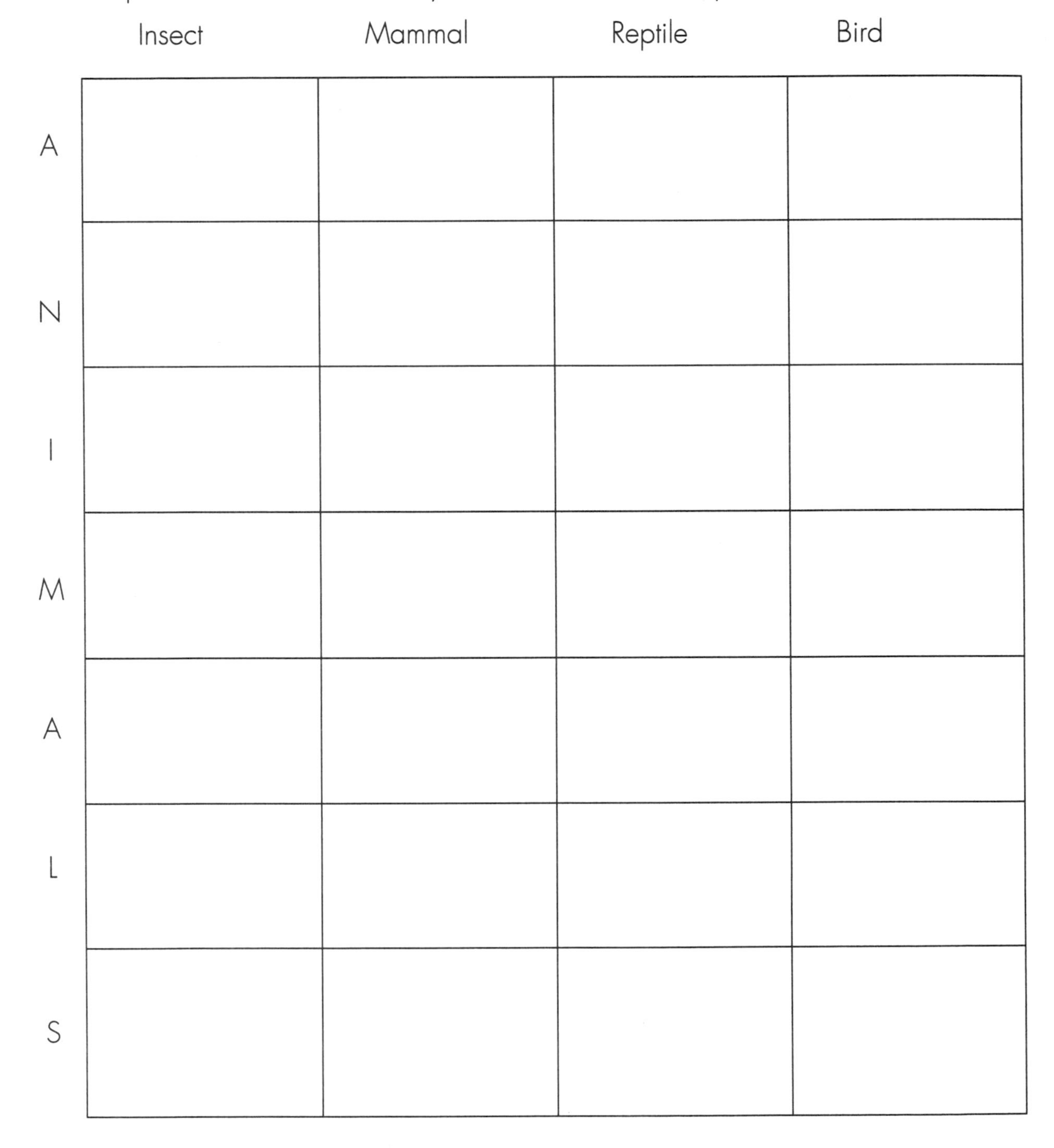

| | Insect | Mammal | Reptile | Bird |
|---|---|---|---|---|
| A | | | | |
| N | | | | |
| I | | | | |
| M | | | | |
| A | | | | |
| L | | | | |
| S | | | | |

Name ______________________________

# Group Therapy

Animals keep company in groups with assorted names. For example, a group of fish is called a *school*. A group of cranes is known as a *sedge* or a *seige*.

Hidden in the story below is a name used for each group of animals listed at the bottom of the page. Find and circle each group name and write it in the correct blank. Words can be hidden within a word or a group of words.

Example: The police guessed gems were stolen. Answer: sedge

Story:

"I play the wrong chord every time!" the star musician said broodingly.
Jeff locked away his guitar in disgust. Turning to the drum set, he played the cymbal, easing into a steady beat.
"I must erase that measure and rewrite it," he thought to himself.
Just then the door flew open.
"How's my husband?" asked Niccol.
"You startled me! I feel like I've seen a ghost! At least ring the bell next time as you enter, dear."
"I'm not too late, am I, to get to the concert tonight? Look at my new gold, glittery dress!" said Niccol, opening a package. Suddenly she looked around.
"Didn't I clean this music alcove yesterday? What has been going on in here?" she asked.
Jeff rubbed his forehead. "Look, not everyone is born neat. This space is mine. Still, I appreciated your help, Niccol, on yesterday's cleaning. Here, have this warm coffee and just relax. I'll take a cab early to the concert on my own. You're all set up. Ride with Jake Trooper and the gang and I'll meet you there."

Animals:

1. ants ____________
2. bees ____________
3. chicks ____________
4. clams ____________
5. ducks ____________
6. elks ____________
7. geese ____________
8. gnats ____________
9. gorillas ____________
10. lions ____________
11. partridges ____________
12. peacocks ____________
13. pheasants ____________
14. pigs ____________
15. ponies ____________
16. sparrows ____________
17. toads ____________
18. turtles ____________
19. wolves ____________
20. kangaroos ____________

Name ____________________

# Label Liabilities

You probably know that in the world of chickens, a male is a *rooster*, a female is a *hen*, and a young offspring is a *chick*. How well do you know the labels for other animals? Below are three choices for each selected animal. Circle the letter in front of the correct answer. Labels are listed in this order: male, female, young.

1. Bear: A) bull, cow, cub B) boar, sow, cub C) buck, doe, cub
2. Cat: A) tom, dame, kitten B) jack, jenny, kitten C) tom, queen, kitten
3. Cattle: A) bull, cow, calf B) cow, cow, calf C) cow, sow, calf
4. Deer: A) dog, doe, fawn B) mare, cow, fawn C) buck, doe, fawn
5. Duck: A) cob, pen, duckling B) drake, duck, duckling C) cob, duck, duckling
6. Elephant: A) boar, cow, pup B) bull, boar, cub C) bull, cow, calf
7. Fox: A) dog, vixen, cub B) buck, doe, pup C) boar, sow, cub
8. Horse: A) mare, doe, pony B) stallion, mare, foal C) mare, stallion, foal
9. Rabbit: A) tom, queen, bunny B) buck, doe, kit C) jack, jenny, bunny
10. Sheep: A) ram, ewe, lamb B) ewe, ram, lamb C) stallion, vixen, lamb
11. Swan: A) gander, swan, duckling B) gander, goose, gosling C) cob, pen, cygnet
12. Swine: A) boar, pig, piglet B) boar, sow, piglet C) bull, sow, piglet
13. Tiger: A) tiger, tigress, cub B) tiger, tigress, foal C) tiger, tigret, cub
14. Whale: A) buck, doe, cub B) boar, sow, calf C) bull, cow, calf

Name ______________________________

# Mammal Scramble

A. Circle the six animals in this list that are mammals, and then try to fit them into the puzzle box. Five animals will go in the horizontal rows and the sixth one will appear diagonally in the circled spaces. Look up any unfamiliar animals in a reference book.

koala vireo
conch hyena
shark sheep
hyrax swift
skate skunk
human nymph

B. Now find the seven mammals in this list and fit them into this second puzzle box. Again, six names will go in the horizontal rows and the seventh will appear diagonally.

baboon marmot
parrot puffer
marten iguana
lizard donkey
turkey kitten
musk ox ferret
oyster plover

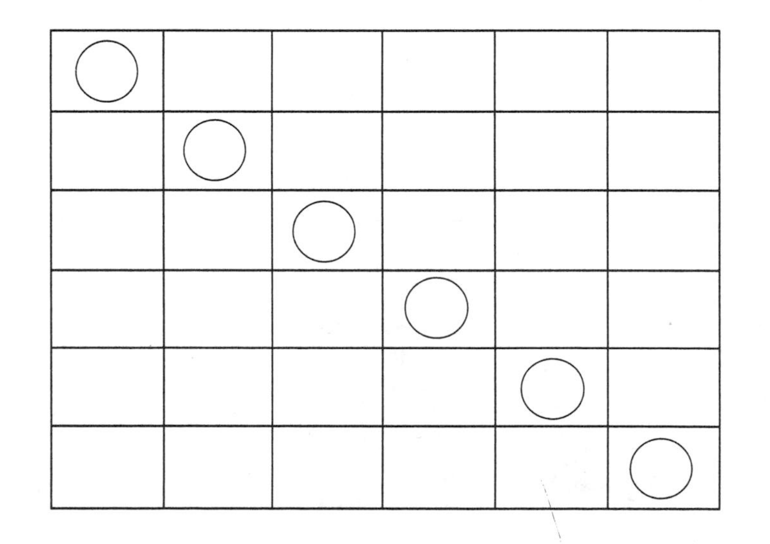

Name ____________________

# Places, Please

Place animals from the word list into the diagrams. All reptiles will fit in the lines that form the word REPTILE, and so on. You will notice that there are more animals than places; that means some do not belong to any of the categories shown.

Word List:

| | | | | | |
|---|---|---|---|---|---|
| abalone | anole | bowfin | cobia | conch | perch |
| gecko | grebe | limpet | lizard | mantid | mosquito |
| murre | mussel | periwinkle | scallop | skink | snail |
| snipe | snook | stinkpot | sulphur | tegula | terrapin |
| veery | whiptail | | | | |

1. _ _ _ R _ _ _ _
2. _ E _ _ _
3. _ _ _ _ _ P _ _
4. _ _ _ _ T _ _ _
5. _ _ I _ _
6. L _ _ _ _ _
7. _ _ _ _ E

8. _ _ _ F _ _
9. _ _ _ I _
10. S _ _ _ _
11. _ _ _ _ H

12. M _ _ _ _ _
13. _ O _ _ _
14. _ _ _ L _ _ _
15. L _ _ _ _ _
16. _ _ _ U _ _
17. S _ _ _ _
18. _ _ _ _ _ _ _ K _ _
19. S _ _ _ _ _ _

20. Into what two categories could your "leftover" animals be placed?
__________ and __________

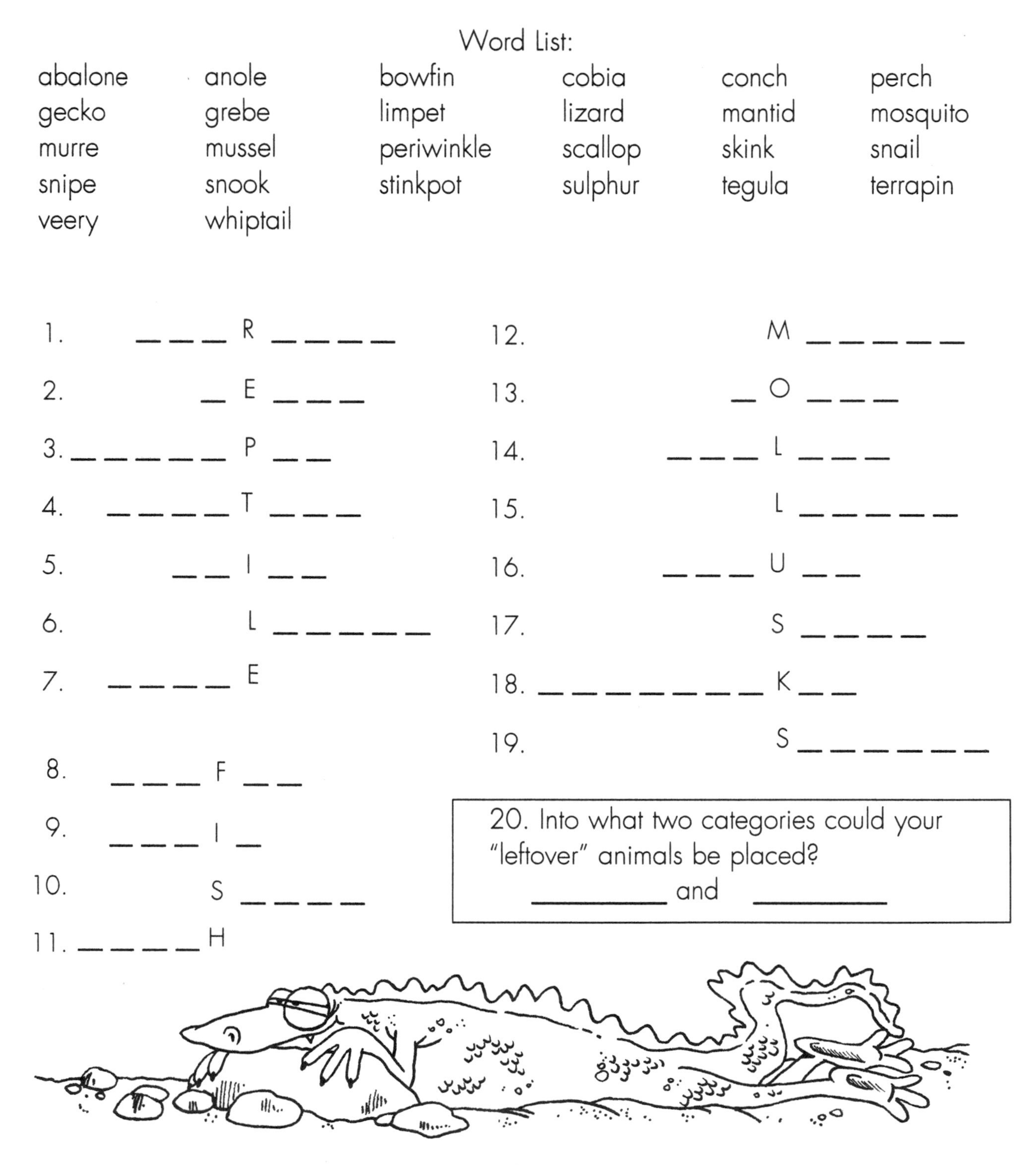

Name ____________________

# Would You Rather Be a Bee?

Study the lives of ants and bees. Learn about their life cycles, their work habits, and the way they cooperate among themselves. Then complete these exercises:

A. List at least four similarities between ants and bees.

1. ____________________
2. ____________________
3. ____________________
4. ____________________

B. List at least four differences between ants and bees.

1. ____________________
2. ____________________
3. ____________________
4. ____________________

C. If you could spend one day as either an ant or a bee, which would you choose? Why?

____________________
____________________
____________________

D. Write a journal entry for the day you became that insect. Tell what you did and where you went. What did you eat? Smell? See? Feel? Hear? Continue your entry on the back of this page.

____________________
____________________
____________________

Name ______________________________

# Think Plants!

As quickly as possible, fill in as many boxes below as you can. For each letter, write the name of a tree, vegetable, fruit, and flower that begins with that letter. For example, in the first box you may choose to write *poplar*. It may not be possible to fill every box. If you can complete all but two or three, you have done extremely well.

| | Tree | Vegetable | Fruit | Flower |
|---|---|---|---|---|
| P | | | | |
| L | | | | |
| A | | | | |
| N | | | | |
| T | | | | |
| S | | | | |

Name ________________________

# A to Z Plants

Place a different letter of the alphabet into each blank to complete the hidden name of a plant. You will not need all the letters in the row to spell a plant name. The lines may spell the name of a tree, wildflower, vegetable, weed, or any other kind of plant. Hint: Cross out each letter of the alphabet as you use it. An example has been completed for you. By adding the letter A to the first line, the word MAPLE is completed. Then circle the name of the plant.

~~A~~ B C D E F G H I J K L M N O P Q R S T U V W X Y Z

1. C H I R M A P L E R T
2. G L A C Y _ R E S S A
3. S T O O P _ O S E N T
4. M A I F O _ G L O V E
5. H U M R A _ I S H A T
6. O M A R I _ O L D E R
7. T R U L L _ C H E N D
8. C R O N A _ O S S U P
9. B A M I L _ W E E D Y
10. W I S H A _ T H O R N
11. M O N E T _ L A X I D
12. A M Y R T _ E T S A Y
13. G E T H U _ O N Q U I L
14. P O O S H _ A T H E R
15. U P C O R _ U I N E L
16. G N I A S _ E R L O Y
17. B U R M A _ A L E A M
18. S T O R C _ I D D L E
19. C U P O L _ I O L E T
20. M A N I L _ C U S T I
21. S Y B R Y _ C C A T T
22. A S P R U _ E A R N D
23. S P A N S _ T E R I A
24. R I P S E _ U O I A N
25. A B L E T _ I R C H Y
26. D R Y N A _ U M A C

Name ______________________________

# Plantegories

All these plants need to be placed into the five categories below. Some plants may appear to belong to more than one category, but there is an arrangement in which each plant will have a place. Look up any unfamiliar plants in a resource book. Write your final answers in the blanks.

Plants:

| | | | |
|---|---|---|---|
| bindweed | bladderwort | bracken | cat briar |
| Douglas fir | dulse | hart's tongue | hemlock |
| honeysuckle | kelp | peas | piñon |
| pitcher plant | rockweed | spleenwort | sugar wrack |
| sundew | tamarack | Venus flytrap | wall rue |

A. Carnivorous (meat-eating) plants: ______________ ______________

______________ ______________

B. Climbing plants: ______________ ______________

______________ ______________

C. Cone-bearing plants: ______________ ______________

______________ ______________

D. Seaweeds: ______________ ______________

______________ ______________

E. Ferns: ______________ ______________

______________ ______________

Name ______________________________

# Fractured Trees

Each description below containing fractions represents the name of a tree. Your job is to decode each tree name by solving the description. It will give you clues to key words. The fraction shows you many letters from the key word you need in the name of the tree.

Example: 2/3 of paddle + 1/4 of save
Answer: The key words are *oar* and *keep*. Use the first 2/3 of oar or *oa*.
Use the first 1/4 of keep, or *k*. Put them together to spell *oak*.

Note: The fractions always refer to the *beginning* letters in the key words.

| | Key Words | Answer |
|---|---|---|
| 1. 2/3 of monkey + 3/6 of promise | ________________ | ________ |
| 2. 3/4 of fowl + 2/4 of buddy | ________________ | ________ |
| 3. 2/3 of energy + 3/6 of wax with a wick | ________________ | ________ |
| 4. 2/3 of gamble + 3/4 of repeating sound | ________________ | ________ |
| 5. 2/5 of arm joint + 1/4 of dairy liquid | ________________ | ________ |
| 6. 2/3 of angry + 3/4 of request | ________________ | ________ |
| 7. 2/5 of sweetener + 3/8 of pasta | ________________ | ________ |
| 8. 2/3 of not high + 4/7 of baked milk and egg dessert | ________________ | ________ |
| 9. 2/3 of pan + 3/4 of scheme + 1/3 of rodent | ________________ | ________ |
| 10. 2/5 of coil + 2/3 of carpet + 2/4 of penny | ________________ | ________ |
| 11. 3/4 of pace + 2/6 of lump + 1/4 of undertaking | ________________ | ________ |
| 12. 3/8 of periodical + 2/5 of sound + 3/4 of person who tells untruths | ________________ | ________ |

Name ____________________

# Mushroom Mania

How well do you know your mushrooms? Take a trip through this maze and find out. Move from START to FINISH only along paths in which you encounter names of genuine North American mushrooms. Mixed in this maze are names of other living organisms, but make certain that your route takes you only past *mushrooms*.

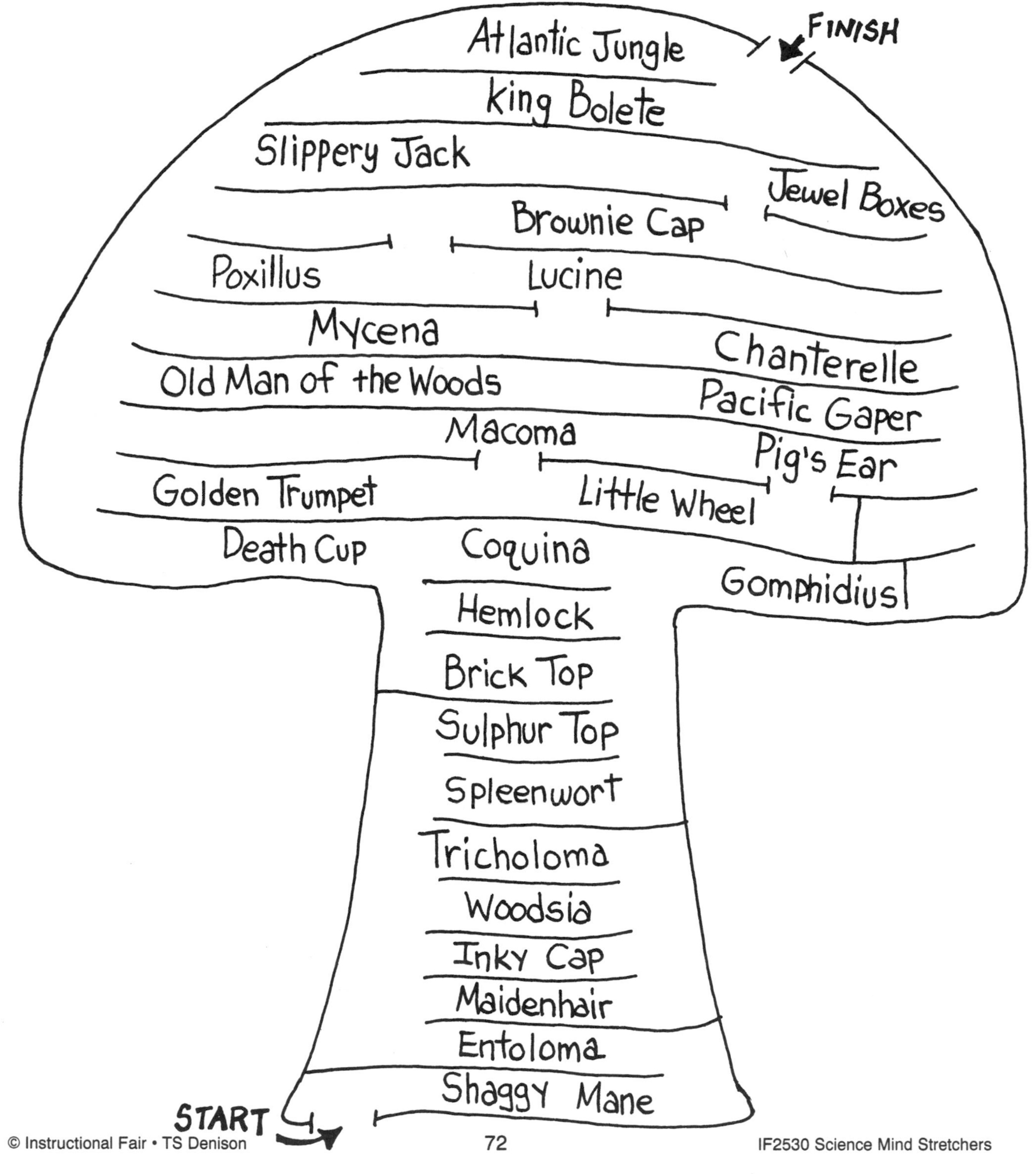

Name ____________________

# Top Ten

Make a "top ten" list of plants—plants that you think are most important to life on earth. Write the name of each plant and one to two sentences about why each is so vital. Draw a picture of one of your top ten plants on the back of this paper.

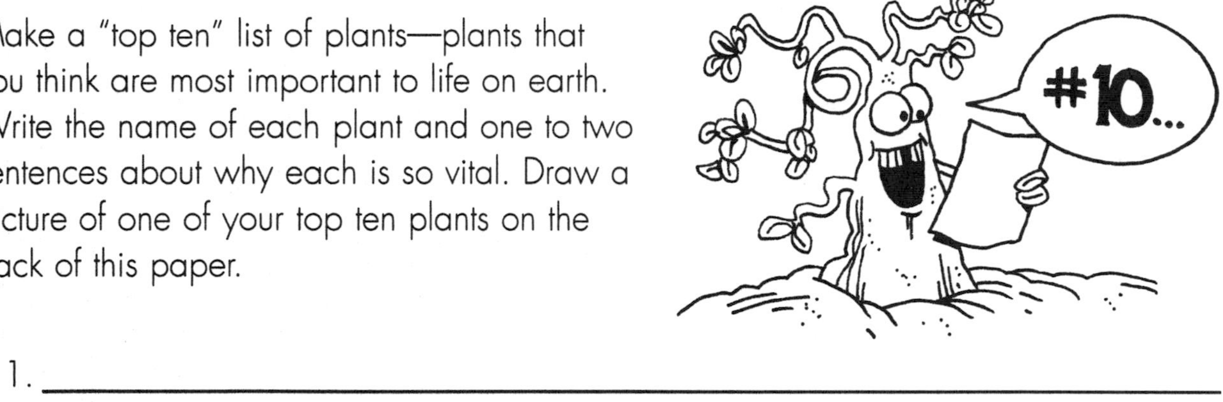


1. ____________________

____________________

2. ____________________

____________________

3. ____________________

____________________

4. ____________________

____________________

5. ____________________

____________________

6. ____________________

____________________

7. ____________________

____________________

8. ____________________

____________________

9. ____________________

____________________

10. ____________________

____________________

# 6 The Human Body

The focus in this section is on the major organs and systems in the body and their functions. Students will become familiar with the names of many organs and the systems to which they belong. They will learn the names of many bones and some of the amazing numbers in the human anatomy. (For example, there are 120 million cells in the retina of one eye alone!) They will also study what organs are affected by certain diseases and explore the value of their five senses.

It will be helpful to have anatomy and health books on hand for student use while they are solving these mind stretchers. Always encourage students to solve as much of the page with the knowledge they already have before consulting the reference books. It should be noted that in these few pages not all the major systems and organs are covered. These activities, however, will spark ideas for other teacher-written or student-written exercises in other related topics.

Warm-up activities:

A. Review (or teach) the definitions of *organ* (a part of the body with a specific function) and *system* (the body's main working units, where many organs function together).

B. Brainstorm as a class to see how many body systems can be listed. Can students also describe the job of each system and some of the organs within each system?

C. Encourage students to list diseases that can affect some or all of the systems listed.

D. Discuss what it would be like to go without each of the five senses, one at a time, for a day. Help students to appreciate the marvel of their senses.

## *Suggestions for Specific Puzzlers*

### Odd Anatomy

Students should be able to recognize some sets quite easily; for others, they may need to look up some of the terms in a good dictionary or an anatomy book. "Odd" ones will not belong because they have a different function or are in a different location from the others.

### Get "Organ" -ized!

This is a fun puzzler where vocabulary is important. Allow students to try it individually at first; later, if some are having problems, have the entire class list possible organs on the chalkboard.

**Anatomical Numbers**

Students should be able to solve most of the lower numbers easily on their own, although even here they may learn something new about the lungs. For the larger numbers, students will have to combine research with guesswork, depending on the resources available. You may wish to allow students to work on this one in pairs.

**Sickly Sleuthwork**

Be certain that students understand the directions—that there is not a one-to-one match-up of letters and numbers. See how many answers students can provide on their own before turning to encyclopedias or dictionaries. As a follow-up, suggest that students select one of the diseases listed on which to write a one- or two-page report. What other diseases can they name that affect the same organs and systems?

**Digging for Bones**

If students need more help in finding bones, you may wish to include a list of the words to be found. You could present them in random order and still require students to place them under the correct headings after they have found the words.

**Come to Your Senses!**

Have students research the function for any of the parts that are unknown to them. Ask students to write their own scrambled words for the other two senses.

**Fitful Fit-In**

This can give anyone fits! Allow plenty of time for students to complete this activity. If the clues provided are not enough, you may have to place letters from the box into the words for them. For example, you could begin by placing all the letters L or C as additional hints.

**Sensitivity Check**

This activity is designed to increase students' awareness and appreciation of their senses. Students may wish to share their writings with the class. Additional related activities could include a study of Helen Keller or a visit to the classroom by someone who is blind or deaf, etc., and is comfortable discussing it with students.

Name ______________________________

# Odd Anatomy

Each group of words below contains one "oddball"—one name that does not belong with the others for some reason. In each line, first cross out the odd part and then in the blank tell briefly why it does not belong.

1. tricep, scapula, patella, clavicle

______________________________

2. bronchial tube, trachea, pancreas, diaphragm

______________________________

3. plasma, aorta, red blood cells, platelets

______________________________

4. kidney, ureter, bladder, lungs

______________________________

5. nose, skin, heart, ear

______________________________

6. medulla, cortex, thalamus, cochlea

______________________________

7. hammer, cornea, anvil, stirrup

______________________________

8. carotid, jugular, atrium, pulmonary

______________________________

9. blood, adrenaline, insulin, thyroxine

______________________________

10. radius, femur, ulna, humerus

______________________________

11. membrane, hypothalamus, nucleus, ribosomes

______________________________

12. muscle, nerve, connective, spine

______________________________

Name ______________________________

# Get "Organ" -ized!

In each box below write the name of a different organ of the human body. When you select the correct organ, you will spell sensible 3-letter words going down each column of the box. Remember that an *organ* is a part of the body that has a specific job to do. Hint: One of the completed vertical 3-letter words in each box is the name of a body part.

Example:

| A | A | P | I | S | R | S |
|---|---|---|---|---|---|---|
| T | R | A | C | H | E | A |
| E | M | D | Y | Y | D | P |

1.

| S | B | E | A | I |
|---|---|---|---|---|
| | | | | |
| E | T | R | T | S |

2.

| O | R | I | B | I |
|---|---|---|---|---|
| | | | | |
| D | B | Y | D | E |

3.

| A | J | A | I | D | J | B | U |
|---|---|---|---|---|---|---|---|
| | | | | | | | |
| E | W | D | E | Y | T | G | E |

4.

| L | D | R |
|---|---|---|
| | | |
| G | E | D |

5.

| E | P | M | L | A |
|---|---|---|---|---|
| | | | | |
| B | Y | P | P | Y |

6.

| E | A | M | O | A | S | A |
|---|---|---|---|---|---|---|
| | | | | | | |
| B | L | N | D | D | T | M |

7.

| S | P | O | I | L | B |
|---|---|---|---|---|---|
| | | | | | |
| Y | N | D | N | G | E |

8.

| A | T | A | E | B | L |
|---|---|---|---|---|---|
| | | | | | |
| E | E | T | G | S | T |

Name ______________________________

# Anatomical Numbers

The human body is truly amazing, and the numbers prove it. Do you know how many bones, muscles, or even cells you have? Try to find out as you complete this page. Select a number from the box to go in each blank below.

| 2 | 3 | 4 | 5 | 22 | 27 | 30 | 32 | 70 | 206 | 500 | 600+ |
|---|---|---|---|---|---|---|---|---|---|---|---|
| 10,000 | | 120 million | | | 700 million | | 30 billion | | 100 trillion | | |

1. Total number of bones in adults ____________________
2. Number of bones in the skull alone ____________________
3. Number of bones in the human hand, the most flexible part of the skeleton ______
4. Number of adult permanent teeth ____________________
5. Number of chambers in the heart ____________________
6. Average number of times an adult's heart beats per minute ____________________
7. Number of red blood cells (the single most abundant cell in the body) in a normal adult ____________________
8. The number of major sense organs ____________________
9. The number of taste buds in the average person's mouth ____________________
10. The number of cells in the retina of an eye ____________________
11. The number of nerve cells in the brain ____________________
12. The approximate number of functions performed by the liver, more than by any other organ ____________________
13. The number of voluntary muscles in the body ____________________
14. The number of lobes in the right lung ____________, left lung ________________
15. Total number of cells in an adult human body ____________________

Name ______________________________

# Sickly Sleuthwork

You have probably heard of most of the diseases listed below, but how much do you know about them? In the blank next to each disease, write the letter that tells what part(s) of the body this disease primarily affects. Some letters will be used more than once. Two letters will not be used at all.

____ 1. multiple sclerosis
____ 2. tuberculosis
____ 3. influenza
____ 4. AIDS
____ 5. cholera
____ 6. impetigo
____ 7. arthritis
____ 8. leukemia
____ 9. colitis
____ 10. glaucoma
____ 11. rheumatic fever
____ 12. hepatitis
____ 13. osteoporosis
____ 14. shingles
____ 15. diabetes
____ 16. dysentery
____ 17. goiter

A. kidney
C. eye
D. white blood cells
E. respiratory system (throat, lungs, etc.)
L. ear
N. thyroid gland
O. digestive system (stomach, intestines, pancreas, liver, etc.)
R. skeletal system (bones, joints, etc.)
S. central nervous system (brain, nerves, spinal cord)
T. heart
U. skin
Y. immune system

If you have solved this puzzle correctly, you will be able to read a piece of medical advice should you contract any alarming symptoms!

Name ______________________________

# Digging for Bones

There are 206 bones in an adult's body, but only 25 of them are hidden here. Locate the names of bones in the word search and circle them. Also, try to find four types of joints that connect bones in the human body. The words may run vertically, horizontally, or diagonally, both backwards and forwards. After you have found the name of a bone or joint and circled it, then write it under the correct heading on the next page. You may need to refer to an anatomy book for extra help. Start digging!

M A X I L L A R O P M E T C

S U I D A R G H I N G E A A

L A T N O R F Y U S K T R I

A S D L A T I P I C C O S B

P M T D A P I V O T E A A I

R L F E L O L S R I L R L T

A A F B R E D A U A C B U D

C T I I P N L I M N I E S I

A E B R A L U Y E L V T A O

T I U L E N O M F U A R C N

E R L T M A N D I B L E R E

M A A L U P A C S I C V U H

B P S L A S R A T A T E M P

C A R P A L S U R E M U H S

Joints:

1.

2.

3.

4.

Bones in the chest and trunk:

1.

2.

3.

4.

5.

6.

7.

Bones in the arms and hands:

1.

2.

3.

4.

5.

Bones in the head:

1.

2.

3.

4.

5.

6.

7.

Bones in the legs and feet:

1.

2.

3.

4.

5.

6.

Name ____________________

# Come to Your Senses!

The human body uses a complex combination of specialized parts in order to see, smell, and hear. For each of these three senses, several body parts have been scrambled. Unscramble each one and write it correctly in the blank provided.

SIGHT:

1. S I I R ____________
2. R E C O N A ____________
3. N A T I R E ____________
4. L I P P U ____________
5. A E F V O ____________
   N E T R C L I S A ____________
6. C L E A R S ____________
7. D I E L E Y ____________
8. I P O C T  N E V E R ____________
9. S C E O N ____________
10. D R O S ____________

HEARING:

11. A R M U D E R ____________
12. M E R M A H ____________
13. V I L A N ____________
14. A L C E H O C ____________
15. N A N I P ____________
16. P R U S R I T ____________

SMELL:

17. M U S T E P ____________
18. A I L I C ____________
19. A C H E N C O ____________
20. L I N S O R T ____________

Name ______________________________

# Fitful Fit-In

Fit each consonant from the box into a blank in the words below. Cross out a consonant each time you use one. When completed, each word will spell the name of a body part.

- Hint #1: The words listed are in alphabetical order.
- Hint #2: Each body part belongs to one of these six systems and is labeled with the corresponding initial:

| | |
|---|---|
| Circulatory - C | Nervous - N |
| Digestive - D | Respiratory - R |
| Muscular - M | Skeletal - S |

B B C C C C C C D D D G G G G H H H H H H J K L L L L L L M M M M M

N N N N N N N N N N N P P P P P Q R R R R R R R R R R S S S S S S S

S S S S S S T T T T T T T T T V V V Y

1. A __ __ E __ __ (C)
2. __ I __ E __ __ (M)
3. __ __ A I __ (N)
4. __ __ A __ I __ __ E (S)
5. __ I A __ __ __ A __ __ (R)
6. E __ O __ __ A __ U __ (D)
7. __ A __ __ __ __ __ I __ __ (M)
8. __ E A __ __ (C)
9. I __ __ E __ __ I __ E __ (D)
10. __ O I __ __ (S)
11. __ U __ __ (R)
12. __ O U __ __ (D)
13. __ E __ __ E __ (N)
14. __ O __ E (R)
15. __ U A __ __ I __ E __ __ (M)
16. __ __ U __ __ (S)
17. __ __ I __ A __ __ O __ __ (N)
18. __ __ E __ __ U __ (S)
19. __ __ O __ A __ __ (D)
20. __ E I __ (C)

Name ______________________________

# Sensitivity Check

List your five senses in the boxes below. Under each sense, list four reasons why that sense is important or enjoyable for you. Be specific. For example, under sight you may write "watching a sunset" or "looking out for traffic when crossing the street."

| |
|---|
| |

1. ______________________
2. ______________________
3. ______________________
4. ______________________

| |
|---|
| |

1. ______________________
2. ______________________
3. ______________________
4. ______________________

| |
|---|
| |

1. ______________________
2. ______________________
3. ______________________
4. ______________________

| |
|---|
| |

1. ______________________
2. ______________________
3. ______________________
4. ______________________

| |
|---|
| |

1. ______________________
2. ______________________
3. ______________________
4. ______________________

Which sense do you think is the most important? ______________________

Why? ______________________

______________________________

If you suddenly lost this sense, what would be your biggest disappointment?

______________________________

______________________________

______________________________

Name ______________________________

# Sensitivity Check II

Write a one-paragraph description for any three of the topics listed below. Write one paragraph on the lines provided and write the other two on the back of this page.

1. Describe eating a chocolate ice-cream cone without the sense of taste.
2. Describe a traffic jam without the sense of hearing.
3. Describe being outside on a cold winter day without the sense of touch.
4. Describe Niagara Falls without the sense of sight.
5. Describe a red rosebud without the sense of smell.
6. Describe a pan of bubbly, hot lasagna without the sense of smell or taste.
7. Describe a beautiful spring day without the sense of sight or hearing.
8. Describe the ocean without the sense of sight.
9. Describe a kitten without the sense of touch.
10. Describe a blazing campfire without the sense of hearing or smell.
11. Describe typing a letter without the sense of touch.
12. Describe homemade chocolate chip cookies without the sense of taste or smell.

Paragraph One:

______________________________________________

______________________________________________

______________________________________________

______________________________________________

______________________________________________

______________________________________________

______________________________________________

# Answer Key

## INVENTION AND INVENTORS

### Chinese Confusion — page 3
TOOTHBRUSH, 1498

### Criss-Crossed Inventors — page 4
1. Roentgen  2. Volta  4. Nobel
5. Howe  6. Kolff  7. Harington
8. Gutenberg  9. Fermi  10. Bell
11. Morse  12. Faraday  13. Otis
14. Townes  15. Page  16. Truong

### Sequential Sets (Earliest is bold.) — page 5
| | | |
|---|---|---|
| 1. A. 1878 | **B. 1764** | C. 1831 |
| 2. A. 1965 | **B. 1850** | C. 1886 |
| 3. A. 1899 | B. 1908 | **C. 372** |
| 4. **A. 535** | B. 1889 | C. 1883 |
| 5. **A. 3875 BC** | B. 1690 | C. 1859 |
| 6. **A. 1834** | B. 1861 | C. 1884 |
| 7. A. 1802 | **B. 980** | C. 1777 |
| 8. A. 1920 | B. 1895 | **C. 1683** |
| 9. **A.1965** | B. 1972 | C. 1970 |
| 10. A. 1803 | B. 1819 | **C. 1684** |
| 11. **A. 1878** | B. 1906 | C. 1905 |
| 12. A. 1937 | B. 1912 | **C. 1848** |

### International Inventiveness — page 6
Country A: England
1. fire escape
2. megaphone
3. magnifying (glass)
4. power loom
5. rubber band
6. band-saw
7. side saddle

Country B: Germany
8. hourglass (set)
9. wheel chair
10. printing press
11. uranium
12. clarinet
13. accordion
14. toy doll

Country C: USA
15. submarine  16. steamboat  17. burglar alarm

### Enterprising Edison — page 7
1. C  10. B
2. B  11. C
3. B  12. C
4. A  13. A, B, E, F
5. C  14. C
6. A  15. B, C, D, E
7. B*  16. B
8. C  17. C
9. A  *(He sat on the eggs and crushed them!)

### Scientific Match-Up — page 9
1. J  11. O
2. F  12. T
3. M  13. L
4. A  14. E
5. P  15. N
6. B  16. Q
7. I  17. K
8. G  18. R
9. S  19. D
10. H  20. C

### "Like"-ly Inventions (Answers may vary) — page 10
1. All were invented by Benjamin Franklin.
2. All were invented in the year 1908.
3. All were invented in the USA in the year 1937.
4. All were for the same purpose of written communication.
5. All were invented in Italy.
6. All were named after their inventors.
7. All were invented or discovered in the year 1894.
8. All were for agricultural purposes.
9. All were invented in France.

### Preposterous Publicity — page 11
A. Kindergarten, 1832  B. Skateboard, 1966

### True Tales — page 12
1. sweeper  6. tanks
2. waterproof  7. reading lens
3. jigsaw puzzle  8. fire extinguisher
4. revolving door  9. pressure cooker
5. bacteria  10. zipper

## COMPUTERS

### Humanity vs. Technology — page 18
1. C  6. C
2. H  7. H
3. H  8. C
4. C  9. C
5. H  10. H

### Basic Match-Up — page 19
1. D, II  4. E, I
2. A, V  5. C, IV
3. B, III

**Device Dilemma** **page 20**

A = S, B = H, C = U, D = L, E = K, F = A, G = T,
H = R, J = B, K = N, L = P, M = E, N = Y, O = G,
P = D, Q = I, R = M, W = O, Z = C

1. BAR CODES (I)
2. PRINTER (O)
3. SCANNER(I)
4. PLOTTER (O)
5. ICONS (I)
6. MOUSE (I)
7. MUSICAL KEYBOARD (I)
8. DISK (I)
9. SCREEN (O)
10. GRAPHICS TABLET (I)

**Chip, Chip Hooray!** **page 21**

1. 7
2. 2
3. 5
4. 3
5. 9
6. 1
7. 8
8. 10
9. 6
10. 4

**Definite Mix-Up** **page 22**

(Slight variations are possible on several of these.)

1. a number system based on two digits: 0 and 1
2. a binary digit, this is 0 or 1
3. an error in a computer program
4. a group of eight binary digits which represents one unit of data in the computer's memory
5. the control center of the computer which organizes all the other parts inside
6. a program that stores information in such a way that it can be accessed very quickly
7. a round flat magnetic plate on which computer data is stored
8. a chart showing the sequence of steps needed for a computer program
9. all the computer equipment including the computer itself, input, output, and storage devices
10. a measure of data storage equal to 1,024 bytes
11. a measure of data storage equal to 1,024 x 1,024 bytes
12. the chips in the computer where information and instructions are stored in binary code
13. a device which enables computers to communicate with each other using telephone lines
14. an input device which you roll around on a desk to move an on-screen pointer to input commands
15. the part of the computer's memory where data, instructions, and results are stored temporarily
16. the part of the computer's memory containing a permanent storage of instructions
17. computer programs on disk or tape
18. a bug deliberately, but secretly, introduced to a computer or software to cause problems for the user

**Binary Baffler** **page 24**

I. A. 25 B. 6 C. 22 D. 15 E. 36 F. 59

II.
1. = 1 A
2. = 10 B
3. = 11 C
4. = 100 D
5. = 101 E
6. = 110 F
7. = 111 G
8. = 1000 H
9. = 1001 I
10. = 1010 J
11. = 1011 K
12. = 1100 L
13. = 1101 M
14. = 1110 N
15. = 1111 O
16. = 10000 P
17. = 10001 Q
18. = 10010 R
19. = 10011 S
20. = 10100 T
21. = 10101 U
22. = 10110 V
23. = 10111 W
24. = 11000 X
25. = 11001 Y
26. = 11010 Z

III. A. gum B. yes C. vote D. water E. lizard F. squash

**Name That Document!** **page 26**

Answers will vary.

**Field Finesse** **page 27**

Answers will vary.
Just be certain that students are following directions.

**Pumpkin Program Problems** **page 28**

(Accept any reasonable variations)

A. 1. Line 2 should tell robot to go to a new patch; otherwise, it could keep visiting the same one.
2. If there were no ten-inch pumpkins anywhere, it would continue indefinitely to try to find some.

B. 1. Leave home.
2. Go to nearest unvisited pumpkin patch.
3. Ask for a 10-inch pumpkin.
4. If patch has none, go back to line 2. Do this only 3 times.
5. Buy pumpkin.
6. Go home.

C. Sample flowchart:

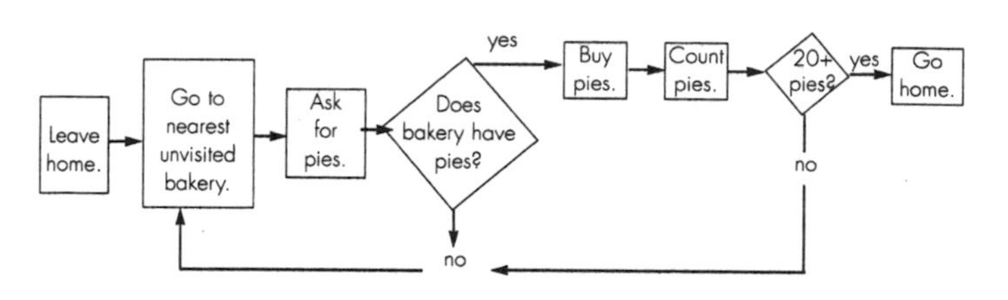

Sample program:
1. Leave home
2. Go to nearest unvisited bakery.
3. Ask for pumpkin pies.
4. If bakery has no pies, go to line 2.
5. If bakery has pies, buy them.
6. Count total number of pies. If less than 20, go to line 2.
7. If total is 20 or greater, go home.

**Keyboard Quest** **page 30**

Answers will vary greatly. Here are some possible outcomes.

A. 4-letter words: crew, ever, free, race, wage
5-letter words: brace, grass, start, waste, zebra
6+-letter words: career, regret Longest: ???

B. 4-letter: jump, kiln, moon, only, pink, pony
5-letter: holly, knoll, milky, plump, nippy
6+-letter: monopoly, lollipop Longest: ???

C. 4-letter: adds, dash, gala, glad, hash
5-letter: flash, flask, glass, salad, shall
6+-letter: ???

D. 4-letter: coal, down, fork, kept, lend, name
5-letter: blame, gland, panel, right, quake, shame
6+-letter: height, turkey, toxicity Longest: ???

## SPACE AND SPACE EXPLORATION

**Planetary Numbers 1** **page 33**

E. Number of natural satellites (moons)
Mars - 3, 4, 3, 687 days
Earth- 1, 3, 5, 365 days
Pluto - 6, 9, 1, 248 years
Venus - 9, 2, 4, 225 days
Saturn - 7, 6, 8, 29 years
Mercury - 4, 1, 2, 88 days
Uranus - 8, 7, 7, 84 years
Neptune - 5, 8, 6, 165 years
Jupiter - 2, 5, 9, 12 years

**Planetary Numbers II** **page 34**

| | |
|---|---|
| 1. 1, 2, 3, 4, 9 | 7. 5, 6, 7, 8 (Pluto is unknown) |
| 2. 5 | 8. 5 |
| 3. 4, 5 | 9. 3 |
| 4. 5, 6, 7, 8 | 10. 5, 6, 7, 8 |
| 5. 2 | 11. 7 |
| 6. 4 | 12. 2, 3, 4 |

**Astronomical Alterations** **page 35**

| | |
|---|---|
| 2. quasar, D | 8. star, J |
| 3. pulsar, F | 9. sun, E |
| 4. orbit, H | 10. comet, G |
| 5. Mars, L | 11. axis, I |
| 6. planet, A | 12. meteor, C |
| 7. moon, B | |

**Solar Calculations** **page 36**

1. 93,000,000 miles
2. A. 1,690,909 hours
   B. 70,455 days
   C. 193 years
   D. 101,454,540 minutes
   E. 6,087,272,400 seconds
3. six billion eighty-seven million, two hundred seventy-two thousand, four hundred
4. A. 163,680,000,000 yards
   B. 491,040,000,000 feet
   C. 149,800,000 kilometers
   D. 149,800,000,000 meters
5. It would take the same amount of time. Distances are all the same; only the units are different.

**Space-Age Phenomena** **page 37**

| | | |
|---|---|---|
| 1. black hole | 6. neutron star | 11. Saturn's rings |
| 2. supernova | 7. solar eclipse | 12. shooting star |
| 3. rainbow | 8. asteroid belt | 13. thunderstorm |
| 4. red giant | 9. northern lights | 14. meteor shower |
| 5. sunspots | 10. white dwarf | 15. solar flares |

**Constellation Conundrum** **page 38**

| | | |
|---|---|---|
| Bird of Paradise | Flying Fish | Air Pump |
| Big Dipper | Water Snake | Serpent-Bearer |
| Southern Fly | Mariner's Compass | Berenice's Hair |
| Northern Crown | Great Dog | Hunting Dogs |
| Sea Serpent | Little Lion | Sculptor's Tool |

**Travel Sequence** **page 39**

| | |
|---|---|
| 1. 8, 1965 | 11. 18, 1986 |
| 2. 15,1976 | 12. 2, 1937-45 |
| 3. 20, 1990 | 13. 12, 1973 |
| 4. 11, 1969 | 14. 19, 1989 |
| 5. 4, 1958 | 15. 6, 1961 |
| 6. 17, 1981 | 16. 14, 1975 |
| 7. 1, 1926 | 17. 3, 1957 |
| 8. 13, 1974 | 18. 7, 1962 |
| 9. 5, 1959 | 19. 9, 1966 |
| 10. 10, 1968 | 20. 16, 1980 |

**Space Search** **page 40**

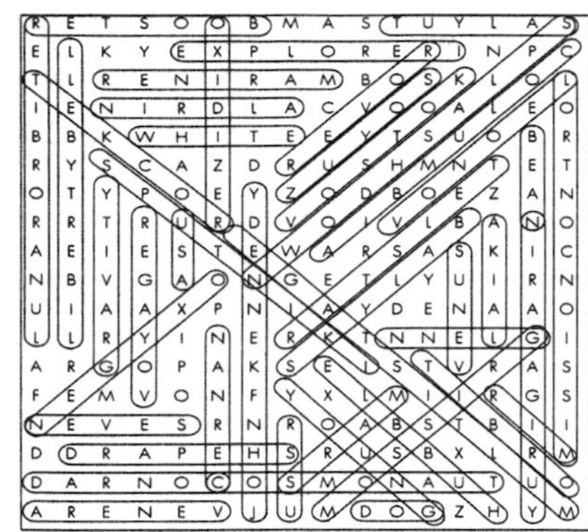

**Space Search** Answers to clues: **page 41**

1. gravity
2. oxygen, oxidizer
3. multi-stage rocket
4. booster
5. splashdown
6. mission control
7. USSR, *Sputnik I*
8. *Explorer I*
9. *Telstar*
10. Laika, Dog
11. Yuri Gagarin, *Vostok*
12. Shepard, orbit
13. Grissom, *Liberty Bell*
14. Glenn
15. Leonov
16. White
17. lunar orbiter
18. John F. Kennedy
19. USA
20. Aldrin
21. Conrad, Bean, Cernan, Young
22. rover
23. *Soyuz*
24. *Mir, Salyut*
25. cosmonaut
26. *Skylab*
27. *Columbia*, seven
28. *Voyager, Mariner*
29. Mars, *Venera*
30. Venus
31. Hubble

**Space History Mysteries** **page 43**

A. Robert Goddard
B. Sally Ride
C. Valentina Tereshkova
D. Robert Gilruth
E. Wernher von Braun

**WEATHER AND CLIMATE**

**Climate Conundrum** **pages 48 and 49**

B. expands 17
C. heavier, 20
D. motion, 25
E. weather, 1
F. directly, 14
G. climate, 4
H. ocean, 30
I. wind, 22
J. vapor, 27
K. pressure, 18
L. period, 6
M. water, 9
N. earth, 2
O. poles, 15
P. rises, 21
Q. blown, 29
R. surface, 24
S. varies, 3
T. equator, 13
U. light, 10
V. angle, 11
W. movement, 16
X. molecules, 19
Y. quarters, 23
Z. evaporates, 26
AA. clouds, 28
BB. pattern, 5
CC. shines, 12
DD. sunlight, 7

**Meteorologist's Mix-Up** **page 50**

1. forecast
2. sunshine
3. rainfall
4. ice storm
5. pressure
6. dense fog
7. blizzard
8. humidity
9. advisory
10. freezing
11. high wind
12. snowfall

**Search High and Low** **page 51**

1. T
2. F - Honolulu
3. M - chart does not show all US cities
4. F - Fairbanks
5. M- Chart does not show day-to-day differences
6. T
7. M - Does not show day-to-day figures
8. T
9. F - the difference is 19 degrees
10. F - the range is 46 degrees
11. M - chart does not show all U.S. cities
12. T
13. M - chart does not show averages for all 12 months
14. F - Fairbanks has a difference of 38 degrees
15. T

**Symbol Sense** **page 52**

A. 6
B. 10
C. 1
D. 16
E. 14
F. 2
G. 13
H. 4 (Symbols 5, 7, 12 and 15 are the "fakes")
I. 9
J. wrong label
K. 11
L. 8
M. 3
N. wrong label

**Where in the World?** **page 53**

1. C
2. H
3. D
4. J
5. L
6. B
7. F
8. A
9. I
10. E
11. K
12. G

**Chilly Challenger** **page 54**

A. 33
B. -5
C. -26
D. -15
E. -3
F. -29
G. -66
H. -56
I. -27
J. -6
K. -46
L. -102

**Temperature Tantrum** **page 55**

(Variations are likely. Numbers used in reaching these solutions are given in parentheses.)

1. 50°F
2. 20°C
3. 7°F (The temp is about 10°F)
4. -31°F (The temp is about 5°F)
5. 15°F
6. -67°F (The wind speed is about 20 mph)
7. -30°F (The wind speed is about 45 mph; the temp is about 15°F)
8. -3°F (The wind speed is 10 mph)
9. 21°F ((The wind speed is 5 mph, the temp is 25°F)
10. -5°F (The wind speed is 30 mph)
11. -20°F (The wind speed is 35 mph)
12. 3
13. Fresh breeze
14. 6 (The temp is about -5°F with 30 mph winds)
15. Moderate gale (The temp is about 30°F)

**Weather Trivia** **page 57**

1. A. troposphere C. ionosphere
   B. stratosphere D. exosphere
2. 23½
3. March 21 and Sept 23
4. equal nights
5. an instrument that measures the speed of wind
6. atmospheric pressure
7. humidity
8. USA, over 500
9. low
10. hurricanes
11. typhoons or cyclones
12. cirrus, cumulus, stratus
13. rain
14. clockwise
15. areas around the equator
16. mid-latitude mountainous regions (such as the Rocky Mountains in the U.S.)

**Weather Wizard II** **page 59**

D. 1946, The method has been used to break up fog at busy airports. In the U.S., it has been used to try to reduce lightning in storms; in Russia, it has been used to reduce the size of damaging hailstones.

## ANIMALS AND PLANTS

**Think Animals!** **page 62**

(Outcomes will vary. Here are some possible answers:)

A - ant, aardvark, alligator, American coot
N - ?, nutria, Northern water snake, nuthatch
I - ichneumon, ?, iguana, ibis
M - mosquito, mink, mud turtle, mallard
A - aphid, anteater, anole, anhinga
L - ladybug, lemming, lizard, lark
S - stink bug, skunk, skink, sparrow

**Group Therapy** **page 63**

Words that should be circled are shown in bold.

"I play the wrong c**hord e**very time!" the star musician said **brood**ingly.

Je**ff lock**ed away his guitar in disgust. Turning to the drum set, he played the cym**bal, e**asing into a steady beat.

"I **must er**ase that measure and rewrite it," he thought to himself.

Just then the door flew open.

"How's my hus**band**?" asked Niccol.

"You startled me! I feel like I've seen a g**host**! At lea**st ring** the bell next time as you enter, dear."

"I'm not too la**te, am** I, to get to the concert tonight? Look at my new gold, g**litter**y dress!" said Niccol, opening a **pack**age. Suddenly she looked around.

"Didn't I clean this music al**cove y**esterday? What has been going on in here?" she asked.

Jeff rub**bed** his forehead. "Loo**k, not** everyone is born neat. This space is mi**ne. St**ill, I appreciated your help, Niccol, on yesterday's cleaning. Here, have thi**s warm** coffee and just relax. I'll take a cab early to the concert on my own. You're all set u**p. Ride** with Jake **Troop**er and the **gang** and I'll meet you there."

1. colony
2. swarm
3. brood
4. bed
5. team
6. gang
7. flock
8. horde
9. band
10. pride
11. covey
12. muster
13. nest
14. litter
15. string
16. host
17. knot
18. bale
19. pack
20. troop

**Label Liabilities** **page 64**

1. B
2. C
3. A
4. C
5. B
6. C
7. A
8. B
9. B
10. A
11. C
12. B
13. A
14. C

## Mammal Scramble — page 65

| h | u | m | a | n |
|---|---|---|---|---|
| h | y | r | a | x |
| s | h | e | e | p |
| s | k | u | n | k |
| k | o | a | l | a |

| m | u | s | k | o | x |
|---|---|---|---|---|---|
| m | a | r | m | o | t |
| f | e | r | r | e | t |
| k | i | t | t | e | n |
| d | o | n | k | e | y |
| b | a | b | o | o | n |

## Places, Please — page 66

1. TERRAPIN
2. GECKO
3. STINKPOT
4. WHIPTAIL
5. SKINK
6. LIZARD
7. ANOLE
8. BOWFIN
9. COBIA
10. SNOOK
11. PERCH
12. MUSSEL
13. CONCH
14. ABALONE
15. LIMPET
16. TEGULA
17. SNAIL
18. PERIWINKLE
19. SCALLOP

20. Into what two categories could your "left-over" animals be placed?
birds and insects

## Think Plants! — page 68

(Outcomes will vary. Here are some possible answers:)
P - poplar, parsnip, pear, peony
L - locust, leek, lemon, larkspur
A - aspen, asparagus, apple, amaryllis
N - Norfolk pine, navy bean, nectarine, narcissus
T - tamarack, turnip, tangerine, tulip
S - sycamore, squash, strawberry, snapdragon

## A to Z Plants — page 69

1. CHIRMAPLERT
2. GLACYPRESSA
3. STOOPROSENT
4. MAIFOXGLOVE
5. HUMRADISHAT
6. OMARIGOLDER
7. TRULLICHEND
8. CRONAMOSSUP
9. BAMILKWEEDY
10. WISHAWTHORN
11. MONETFLAXID
12. AMYRTLETSAY
13. GETHUJONQUIL
14. POOSHEATHER
15. UPCORNUINEL
16. GNIASTERLOY
17. BURMAZALEAM
18. STORCHIDDLE
19. CUPOLVIOLET
20. MANILOCUSTI
21. SYBRYUCCATT
22. ASPRUCEARND
23. SPANSYTERIA
24. RIPSEQUOIAN
25. ABLETBIRCHY
26. DRYNASUMAC

## Plantegories — page 70

A. Venus flytrap, bladderwort, pitcher plant, sundew
B. bindweed, cat briar, honeysuckle, peas
C. Douglas fir, hemlock, piñon, tamarack
D. dulse, kelp, rockweed, sugar wrack
E. bracken, hart's tongue, spleenwort, wall rue

## Fractured Trees — page 71

(Answers, followed by key words)
1. apple (ape, pledge)
2. birch (bird, chum)
3. pecan (pep, candle)
4. beech (bet, echo)
5. elm (elbow, milk)
6. maple (mad, plea)
7. sumac (sugar, macaroni)
8. locust (low, custard)
9. poplar (pot, plan, rat)
10. spruce (spring, rug, cent)
11. walnut (walk, nugget, task)
12. magnolia (magazine, noise, liar)

## Mushroom Mania — page 72

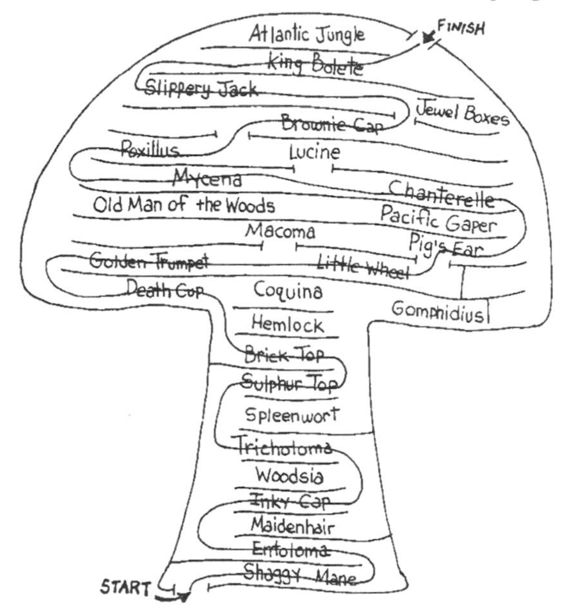


## THE HUMAN BODY
## Odd Anatomy — page 76

1. tricep, it is not a bone
2. pancreas, it is not a part of the respiratory system
3. aorta, it is not a part of the blood
4. lungs, they are not a part of the urinary system
5. heart, it is not a sense organ
6. cochlea, it is not part of the brain
7. cornea, it is not part of the ear
8. atrium, it is not the name of a vein or artery
9. blood, it is not a hormone
10. femur, it is not an arm bone
11. hypothalamus, it is not a part of a cell
12. spine, it is not a type of tissue

## Get "Organ"-ized! page 77

| | |
|---|---|
| 1. heart | 5. brain |
| 2. liver | 6. bladder |
| 3. pancreas | 7. kidney |
| 4. eye | 8. tongue |

## Anatomical Numbers page 78

| | |
|---|---|
| 1. 206 | 8. 5 |
| 2. 22 | 9. 10,000 |
| 3. 27 | 10. 120 million |
| 4. 32 | 11. 30 billion |
| 5. 4 | 12. 500 |
| 6. 70 | 13. 600+ |
| 7. 30 billion | 14. 3, 2 |
| | 15. 100 trillion |

## Sickly Sleuthwork page 79

The letters going down in order spell:
SEE YOUR DOCTOR SOON

## Digging for Bones page 80

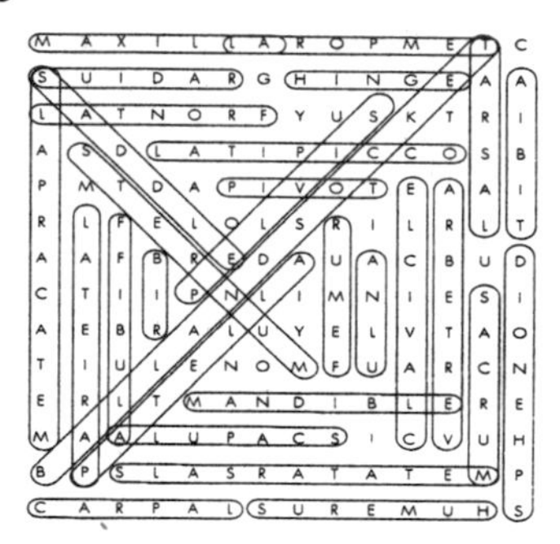

Joints: ball and socket, hinge, saddle, pivot

Bones in the chest and trunk: sternum, rib, clavicle, scapula, pelvis, vertebra, sacrum

Bones in the arms and hands: humerus, ulna, radius, carpal, metacarpals

Bones in the head: frontal, parietal, temporal, occipital, spheroid, maxilla, mandible

Bones in the legs and feet: femur, patella, tibia, fibula, tarsal, metatarsals

## Come to Your Senses! page 82

| | |
|---|---|
| 1. iris | 11. eardrum |
| 2. cornea | 12. hammer |
| 3. retina | 13. anvil |
| 4. pupil | 14. cochlea |
| 5. fovea centralis | 15. pinna |
| 6. sclera | 16. stirrup |
| 7. eyelid | 17. septum |
| 8. optic nerve | 18. cilia |
| 9. cones | 19. conchae |
| 10. rods | 20. nostril |

## Fitful Fit-In page 83

| | |
|---|---|
| 1. artery | 11. lung |
| 2. biceps | 12. mouth |
| 3. brain | 13. nerves |
| 4. clavicle | 14. nose |
| 5. diaphragm | 15. quadriceps |
| 6. esophagus | 16. skull |
| 7. hamstring | 17. spinal cord |
| 8. heart | 18. sternum |
| 9. intestines | 19. stomach |
| 10. joint | 20. vein |